SAE EDGE™
RESEARCH REPORT

Unsettled Issues Concerning the Use of Fuel Cells in Electric Ground Vehicles

Bartlomiej Kolodziejczyk, PhD
H2SG Energy Pte Ltd.

CONTRIBUTORS

Alena Fargere, *World Energy Council*
James Carton, *Dublin City University*
Anna Trendewicz, *E.ON*
Jose Castro Nieto, *Ad Astra Rocket Company*
Craig Knight, *Horizon Fuel Cell Technologies*
Natalia Macauley, *Los Alamos National Laboratory*
Sidarth Komini Babu, *Los Alamos National Laboratory*
Wee-Liat Ong, *Zhejiang University*
David Dvorak, *University of Maine*
Lech Birek, *Shell Energy*

SAE INTERNATIONAL®
Warrendale, Pennsylvania, USA

About the Publisher

SAE International® is a global association of more than 128,000 engineers and related technical experts in the aerospace, automotive, and commercial-vehicle industries. Our core competencies are lifelong learning and voluntary consensus standards development. Visit sae.org.

SAE EDGE™ Research Report Disclaimer

SAE EDGE™ Research Reports focus on topics that are dynamic, in which knowledge is incomplete, and which have yet to be standardized. They represent the collective wisdom of a group of experts and serve as a practical guide to the reader in understanding unsettled subject matter. They are not meant to provide a recommended practice or protocol. The experts engaged have contributed their own thoughts and points of view, and these are not the positions of the institutions or businesses with which they are affiliated. A professional writer has collectivized their input; there is no one contributor's perspective being advanced but rather that of a community of practitioners. SAE EDGE™ Research Reports are the property of SAE International, and SAE alone is responsible for their content.

About This Publication

SAE EDGE™ Research Reports provide state-of-the-art and state-of-industry examinations of the most significant topics in mobility engineering. SAE EDGE™ contributors are experts from research, academia, and industry who have come together to explore and define the most critical advancements, challenges, and future direction in areas such as vehicle automation, unmanned aircraft, Internet of Things (IoT) and connectivity, cybersecurity, advanced propulsion, and advanced manufacturing.

Related Resources

SAE MOBILUS® Power/Powertrain Knowledge Hub
https://saemobilus.sae.org/power/

SAE Team

Frank Menchaca, Chief Product Officer
Michael Thompson, Director, Standards, Information and Research Publications
Monica Nogueira, Acquisitions Director
Jill Leonard, Product Manager
William Kucinski, Managing Technical Editor

Printed in USA

ISSN 2640-3536
e-ISSN 2640-3544
ISBN 978-1-4686-0101-5

To purchase bulk quantities, please contact: SAE Customer Service

E-mail: CustomerService@sae.org
Phone: 877-606-7323 (*inside USA and Canada*)
724-776-4970 (*outside USA*)
Fax: 724-776-0790

Visit the SAE International Bookstore at books.sae.org

About the Series Editor

Bart Kolodziejczyk, PhD, is Chief Technology Officer of the Singapore-based H2SG Energy Pte Ltd. In his role, Kolodziejczyk develops hydrogen generation technologies and works with a number of customers throughout the Asia-Pacific region to implement hydrogen solutions in transportation, energy storage, and other industrial applications. Kolodziejczyk became interested in hydrogen power in 2009 while developing and testing fuel cell inverters for Danfoss Solar Inverters A/S. Kolodziejczyk holds a master's degree in Renewable Energy Science with focus on Fuel Cell Systems and Hydrogen awarded jointly by the University of Iceland and the University of Akureyri, as well as two PhDs in Materials Engineering from Monash University in Australia and Microelectronics from Ecole des Mines de Saint-Etienne in France. Kolodziejczyk has extensive research experience in electrochemistry, catalysis, fuel cell development, and hydrogen generation. He was a Research Fellow at Carnegie Mellon University, where he explored materials for biosensing and energy applications. Kolodziejczyk has advised the United Nations, Organisation for Economic Co-operation and Development (OECD), G20, and European Commission Council of Science, Technology, and Innovation Policy and was named one of MIT Technology Review's Innovators Under 35 for developments of new energy materials and catalysts. Kolodziejczyk has published extensively, covering electrocatalysis and hydrogen generation and its application, among other energy fields. Kolodziejczyk is a Fellow of the Royal Society of Arts, Member of the Royal Society of Chemistry, and Chartered Environmentalist.

contents

Unsettled Issues Concerning the Use of Fuel Cells in Electric Ground Vehicles

Abstract

Hydrogen fuel is rapidly emerging as a clean energy carrier solution that has the potential to decarbonize a variety of industries, including, or predominantly, the transportation industry. Fuel cell electric vehicles (FCEVs), which electrochemically combine stored hydrogen with atmospheric oxygen to efficiently generate electricity while producing only water vapor and small amounts of heat, are heralded to be a game-changing technology. The so-called hydrogen economy has the potential to displace traditional fossil fuel-based economy, with the transportation industry being the first mover in the hydrogen space.

Technological advances made in the last decade in the areas of hydrogen generation and fuel cell technology have enabled the current uptake of hydrogen-based solutions for vehicle applications. Reduced costs, climate change, and carbon tax mechanisms are driving many governments, manufacturers, and consumers toward hydrogen-powered vehicles. The major drawbacks of hydrogen compared to the other competing clean-energy technologies (e.g., battery power), is the high cost of hydrogen refueling and FCEVs. However, application of the economy of scale will enable further cost reduction and broad international uptake of hydrogen in automotive applications.

This Society of Automotive Engineers (SAE) EDGE™ Research Report explores the opportunities and challenges of hydrogen and fuel cell systems in the automotive industry. With the help of expert contributors, several different technological, economic, and safety aspects are considered to develop a better understanding of this emerging hydrogen-based automotive industry. While debates between proponents of battery electric vehicles (BEVs) and FCEVs continue, the current report discusses the unsettled issues in the latter technology and presents a critical overview of the hydrogen and fuel cell systems in the automotive industry. Finally, the report concludes with a series of recommendations aimed at the industry and government stakeholders for implementing and advancing hydrogen transportation projects.

NOTE: SAE EDGE™ Research Reports are intended to identify and illuminate critical issues in emerging, but still unsettled, technologies of interest to the mobility industry. The goal of SAE EDGE™ Research Reports is to stimulate discussion and work in the hope of promoting and speeding resolution of identified issues. SAE EDGE™ Research Reports are not intended to resolve the issues they identify or close any topic to further scrutiny.

BARTLOMIEJ KOLODZIEJCZYK, PhD

H2SG Energy Pte Ltd.

Contributors

Alena Fargere, *World Energy Council*
James Carton, *Dublin City University*
Anna Trendewicz, *E.ON*
Jose Castro Nieto, *Ad Astra Rocket Company*
Craig Knight, *Horizon Fuel Cell Technologies*
Natalia Macauley, *Los Alamos National Laboratory*
Sidarth Komini Babu, *Los Alamos National Laboratory*
Wee-Liat Ong, *Zhejiang University*
David Dvorak, *University of Maine*
Lech Birek, *Shell Energy*

ISSN 2640-3536

Introduction

State of the Industry

The concept of using hydrogen as a clean fuel for transportation is not new. Over the last several decades, many visionaries tried to introduce hydrogen into mainstream mobility. This effort was not purely driven by the vision of one person, but the concept of using clean fuel, potentially derived from water in the electrolysis process, was so convincing that governments and entire nations supported it. For example, Iceland, where the majority of the energy comes from renewable sources such as geothermal and hydropower, is a proponent of hydrogen mobility mainly because of the low cost of renewable energy produced locally and the desire to be completely energy independent from importing fossil fuels for transportation. The first commercial hydrogen refueling station was commissioned by Shell in Iceland's capital Reykjavik in 2003 [1]. Other countries also shared the vision of harnessing this clean fuel for transportation. In the 1960s, the National Aeronautics and Space Administration (NASA) deployed fuel cells in the Gemini and Apollo spacecraft [2]. At that time, fuel cells were less bulky compared to batteries and more efficient that solar panels. They also produced water as a by-product which is useful in space. In the United States, the Energy Policy Act of 1992 was the first national legislation that called for large-scale hydrogen research [3]. The Bush administration had a grand plan to invest $1.2 billion toward developing FCEVs, which would use hydrogen stored onboard to generate electricity for powering the vehicle's electric motor. However, this technology was not mature or cost-competitive enough to compete with cheap fossil fuels and well-established internal combustion engine (ICE) technology. During the Obama administration, the hydrogen program was ended [4] simply because the technology was not ready for widespread commercial adaptation.

In truth, the hydrogen and fuel cell industry required another decade and hundreds of millions in global research funding to make it efficient and cost-competitive. Hydrogen is currently making a comeback, and this time it seems to be here to stay. The renewed interest in hydrogen is driven by a number of factors. First, the technology is more mature and market ready. Second, hydrogen technology is being driven by policy (e.g., the Paris Agreement [5] on climate change), which has enabled research funding and established subsidies for decarbonizing transportation systems which would otherwise be carbon-taxed. These policies have either enabled further funding and subsidies to introduce hydrogen and fuel cells, predominantly to decarbonize the transportation, or enforced hefty carbon taxation schemes, making clean fuels like hydrogen a viable alternative to fossil fuels (Figure 1).

The transition to the hydrogen-based economy is underway, the hydrogen generation technology, as well as fuel cell technology, is probably not entirely market ready, especially cost-wise. However, it is believed that slow, but consistent introduction of these technologies and further hydrogen refueling infrastructure development will enable further cost reduction especially once the economy of scale is met. FCEVs will become cost-competitive to traditional combustion engine counterparts.

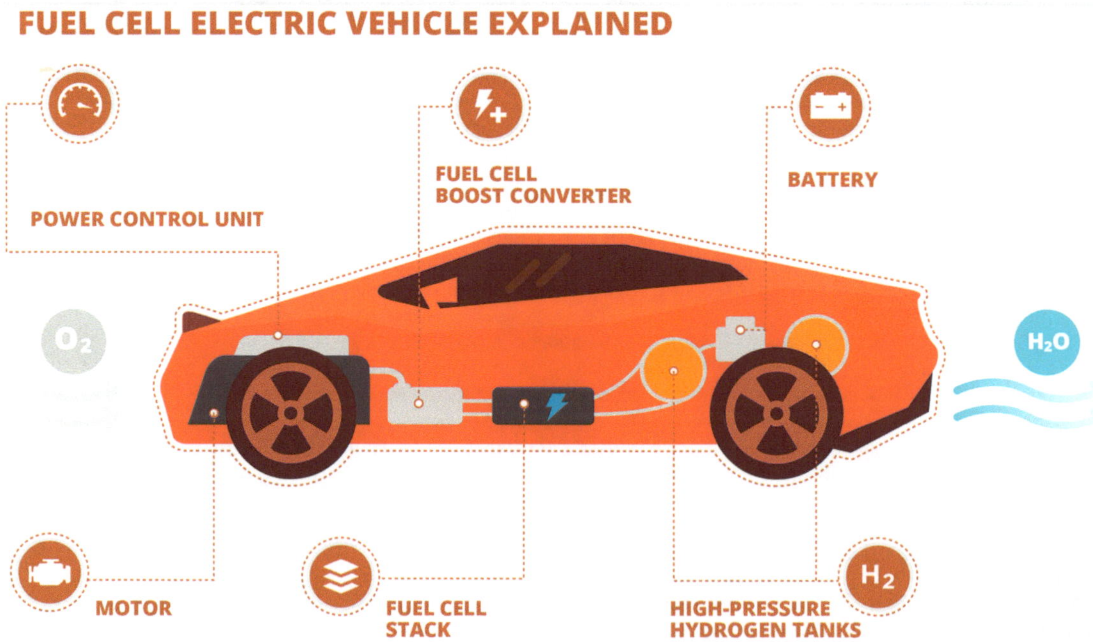

FIGURE 1. FCEV explained. Reproduced from Toyota Europe.

Unsettled Domains in Fuel Cell Electric Vehicle Industry

This SAE EDGE™ report considers the following five areas to be unsettled domains in the fuel cell electric ground vehicles:

- Technology readiness
- Availability of vehicles and infrastructure
- Economic aspects of fuel cell vehicles and hydrogen generation technologies
- Education and social aspects
- Hydrogen safety

While this report covers all five topics, the first three - technology, availability, and economics - are covered in greater depth than education and safety. Readers are encouraged to refer to the provided references or explore the subject further in other professional literature.

The following sections discuss the respective topics, challenges, and potential solutions for each of these areas in more detail. The goal of this SAE EDGE™ Research Report is to educate and provide sufficient perspective on the current state of the industry and to understand potential directions the industry would be heading towards. Finally, this paper will offer some recommendations for the industry and government stakeholders to consider when implementing and advancing hydrogen transportation projects.

Fuel Cell Technology: Past, Present, and Future

Fuel Cell Systems and Their Types

Fuel cells are electrochemical devices that convert the chemical energy of a fuel, most commonly hydrogen, and an oxidizing agent, often oxygen, directly into electricity and heat. Fuel cells are often compared to flow batteries in that they require a continuous flow of fuel and oxidizing agent to generate energy.

While fuel cells are not a new technology (they were invented in 1838 by Sir William Grove), their applications have been limited, mainly because of technology immaturity, high cost, and the lack of a reliable source of hydrogen. The last two decades saw increasing interest in fuel cell applications, spurred mainly by their emission-free operation. These last two decades of intensified research into fuel cells and hydrogen have improved the performance while reducing manufacturing costs. More recently, fuel cells and hydrogen are gaining further momentum and are heralded as a potential alternative to replace fossil fuels for addressing climate change. Some of the most promising applications of fuel cells are in mobility.

Fuel cells comprise three adjacent components - an anode, a cathode, and an electrolyte. Chemical reactions occurring at the anode-electrolyte-cathode interface consume fuel and generate electric current, heat, and water vapor (or carbon dioxide in the case of a direct methanol fuel cell). Electricity generated in the fuel cell can then be applied to power electrical devices, such as the electric motor in an electric vehicle. The electrochemical conversions at the interfaces between the electrodes and the electrolyte are generally efficient, with efficiencies above 65%.

Over the years, several unique types of fuel cells have emerged including proton exchange membrane (PEM), phosphoric acid, molten carbonate, solid oxide, and alkaline fuel cells. While the operational principle is generally the same, operational conditions may differ. For example, solid oxide fuel cells, which use ceramic electrolyte, operate at temperatures between 800 and 1,200°C, whereas PEM fuel cells that rely on a polymeric electrolyte can operate at temperatures below 80°C. The remainder of this report will focus on the application of PEM fuel cells in the automotive industry.

PEM fuel cells have proven to be the most suitable fuel cell type for automotive and portable applications thanks to their lightweight design, high conversion efficiencies, and low operating temperature.

The recent advent of fuel cell applications in the automotive industry allowed further scale-up of PEM fuel cells to systems larger than 100 kilowatts, enabling fuel cell applications in buses, trucks, and other heavy-duty vehicles.

Fuel Cell Electric Vehicles versus Internal Combustion Engine and Battery Electric Vehicles: Technology Comparison

Numerous parameters have to be considered when evaluating FCEVs, BEVs, and ICE vehicles. Some of these parameters are listed in Table 1. Generally, FCEVs and BEVs can be completely emission-free, depending on where the hydrogen and electrical charge come from. Besides, FCEVs have a superior driving range and refueling time compared to battery counterparts. Unfortunately, both BEVs and FCEVs have a limited recharging and refueling infrastructure with medium to high cost of ownership.

Fuel cells are superior technology in terms of energy density, but generally fail to provide high power density (Figure 2). For automotive applications, the most desired energy source would have high power density and high

TABLE 1. Technology comparison.

Parameter	Internal combustion engine vehicle	Battery electric vehicle	Fuel cell electric vehicle
Fuel type	Petrol or diesel	Electricity	Hydrogen
Average fuel economy	10.1 l/100 km	18.4 kWh/100 km	1.3 kg H_2/100 km
Effective cost per kilometer	\$0.071	\$0.041	\$0.058
Greenhouse gas emissions	219-257 g/km	0-133 g/km (depending on electricity origin)	0-226 g/km (depending on hydrogen and electricity origin)
Average driving range	673 km	270 km	472 km
Refueling time	~5 min (light vehicles) Up to 15 min (bus) >15 min (trucks and heavy vehicles)	30 min (DC level 2) Up to 12 hours (AC level 2)	~5 min (light vehicles) Up to 15 min (bus) 10 to 30 min (heavy vehicles)
High voltage (>200 V)	No	Yes	Yes
High pressure (350 to 700 bar)	No	No	Yes
Availability of qualified maintenance personnel	Yes	Limited	Limited
Refueling infrastructure	Commonly available	Limited	Very limited
Cost of ownership[a]	Low to medium	Medium	High

[a] Highly dependent on vehicle type and brand, but also current cost of hydrogen production without economy of scale.

energy density. Currently, only fossil fuel-based ICEs can meet this demand.

As shown in Figure 2, various types of batteries provide similar or better power density compared to fuel cells, but their energy densities are significantly lower than those of fuel cells or ICEs. As such, BEVs typically have a significantly shorter driving range when compared to the other two technologies.

Figure 3 shows the vehicle weight as a function of the vehicle range. Extending the fuel cell vehicle range involves scaling up the hydrogen storage capacity with little impact on the weight. In comparison, the BEV weight increases dramatically for driving ranges greater than 150 to 200 kilometers. Given the lower energy densities of batteries, each additional kilogram of battery to increase the range requires additional structural weight to support the heavier battery, as well as more powerful and bulkier brakes, and a larger electric motor. The resulting compounding weight effectively limits the battery vehicle driving range. There is an optimal tradeoff between battery weight and vehicle range. This issue is nonexistent with FCEVs.

Technology Readiness Level

Technology readiness levels (TRLs) and manufacturing readiness levels (MRLs) for fuel cell systems and FCEVs are very high and growing. While the equipment has reached TRL 9, manufacturing is still limited due to the high technology costs, and vice versa. High technology costs are driven by limited manufacturing capabilities, as well as limited but rapidly growing demand. It is believed that once the economy of scale is reached, technology costs will be significantly reduced.

While technology readiness is high, the relative market immaturity enables technology competition in the areas of hydrogen generation, storage, and utilization. There are several different fuel cell types, each one operating under unique conditions suitable for different applications. PEM fuel cells are the current favorite for automotive applications, thanks to their lightweight design, high efficiencies, and relatively low operating temperatures. While these competing technologies are simultaneously improved, new solutions constantly emerge. The challenge to date is to identify which of the new technologies are the most promising. The complexity and maturity of these different technologies vary significantly across the TRL spectrum with unpredictable prospective costs (since their MRL are often low). Systematic global reviews by energy and science experts can help to identify a list of credible technologies with proven longevity and performance to direct future developments towards these promising solutions. The Office of Energy Efficiency and Renewable Energy at the U.S. Department of Energy has been one of the major voices by setting standards and milestones for fuel cell performance and cost reduction [7].

The major barrier to the deployment of FCEVs is high cost and costs must be reduced across the entire system, from hydrogen generation and fuel cells to hydrogen storage and FCEVs. Generally, the cost of fuel cell raw materials, except for the cost of platinum catalyst, proved not to be a barrier for integrating this technology. Reaching an economy of scale and achieving price competitiveness against competing technologies is the crucial factor for fuel cell vehicle implementation. Increasing and automating fuel cell production is an

FIGURE 2. **Power density and energy density comparison for selected energy storage technologies. Notice the use of a logarithmic scale on both axes.**

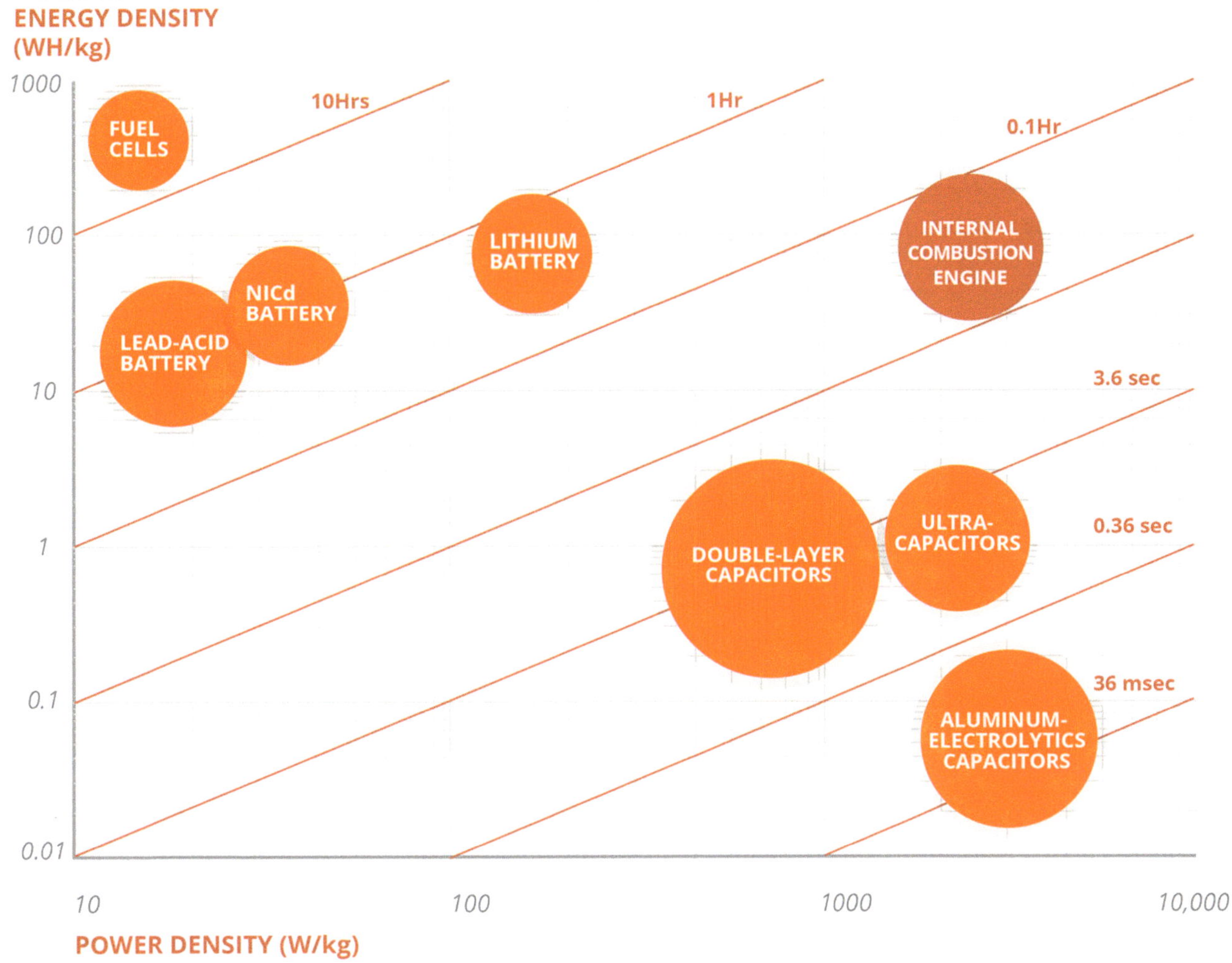

established route to reduce the cost of hydrogen technologies, mirroring the developments in the photovoltaics industry.

Materials research has proved to be an integral part of accelerating market penetration of fuel cell technologies. For example, two decades of ongoing materials research and development allowed significant improvements in PEM fuel cell technology while reducing the technology costs by half through cutting the use of platinum by fivefold and other material improvements [8]. Increased research and development in the field of fuel cells and storage materials will enable and accelerate further technological progress and market uptake by increasing performance and reducing costs, especially for the automotive and portable applications.

In terms of FCEV performance, this has already been validated, mainly in Europe and California with hundreds of vehicles on the roads and a total of several million kilometers driven in real-world conditions [6].

Distinct Segments of the Global Hydrogen Fuel Cell Vehicle Market by Vehicle Type

Strong motivation for decarbonization of the transportation sector is the main driver for the switch towards electric mobility. The variety of vehicles serving different purposes such as for people transport, heavy-duty goods transport, or local warehouse logistics will require different electrification solutions. Therefore, both the BEVs and hydrogen FCEVs are expected to be part of the sustainable energy future. Major original equipment manufacturers (OEMs), including Toyota, Honda, Daimler, etc., have either released

FIGURE 3. Calculated weight of FCEVs and BEVs as a function of the vehicle range. Reproduced from Thomas [6].

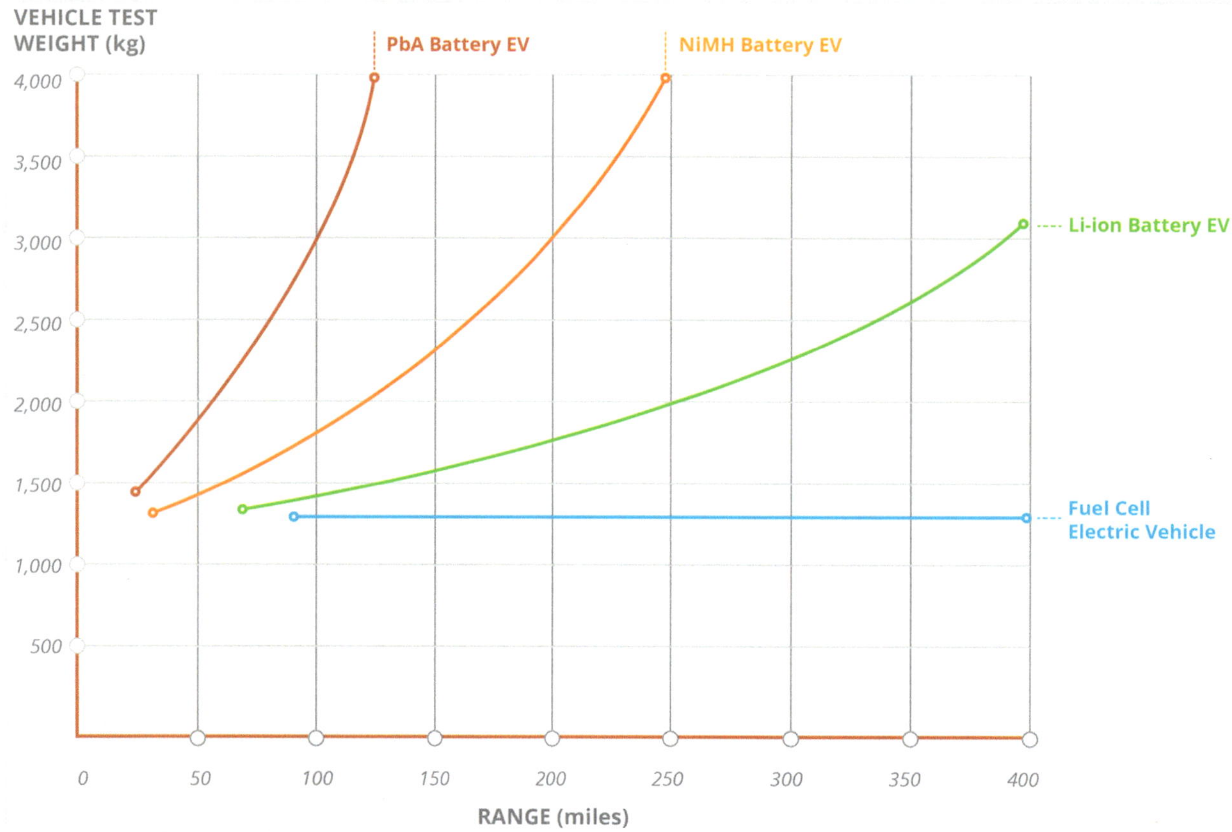

their first model of hydrogen FCEVs or announced plans to do so in the near future.

At the end of 2017, 6,475 FCEVs were registered globally with over half in California, putting the United States (53%) at the forefront for FCEV adoption. Japan takes second place with 38%, while Europe is at 9% [9]. The population of FCEVs in California grew from 331 in 2016 to over 2000 in 2017 and is projected to increase to 13,400 by 2020 [10]. The Northeast Electrochemical Energy Storage Cluster (NEESC), a consortium of eight eastern states in the United States, is aiming to roll out hydrogen refueling stations and up to 10,000 vehicles by 2025 [11].

On the manufacturer side, Toyota is the major player that delivered more than 75% of all the hydrogen fuel cell cars ever made, followed by Honda and Hyundai (13% and 11%, respectively). By 2021, a total of 11 automakers are expected to offer FCEVs (Figure 4).

The following market review briefly summarizes the key global developments of hydrogen fuel cell vehicles by vehicle type.

Passenger Vehicles

Passenger vehicles remain hard to deploy mostly driven by high competition from BEVs and slow refueling infrastructure deployment. While the sales of earlier movers, Toyota Mirai and Honda Clarity continue, 2018 saw the addition of new models that include the Daimler GLC and Hyundai Nexo. Toyota is planning to ramp up its manufacturing to 30,000 FCEVs per annum in 2020, while Hyundai aims to increase production capacity to 40,000 vehicles per annum by 2022. Other manufacturers, including BMW and Volkswagen, also mentioned FCEV activities in their post-2020 visions.

North America accounted for nearly 67% of the global fuel cell cars sales in 2018, concentrated in the state of California with the Toyota Mirai accounting for more than half of "those" sales [13]. The second biggest market with 30% global sales share was in Asia, driven by Hyundai in South Korea and Toyota in Japan. The contribution from Europe was limited in 2018, accounting for just a few percentage points of global sales. While the deployment of FCEVs is still relatively slow, new car manufacturers are continuously entering this new market. Riversimple plans to launch 20 Rasa vehicles in the United Kingdom [14], and Microcab is expected to launch their Vianova model in 2019 [15]. In the high-speed car segment, GreenGT continues to test its newest LMPH2G model that uses a four-stack fuel cell and four electric motors to deliver a combined 653 hp with a top speed of around 300 km/h [16].

FIGURE 4. **Global hydrogen FCEV production [12].**

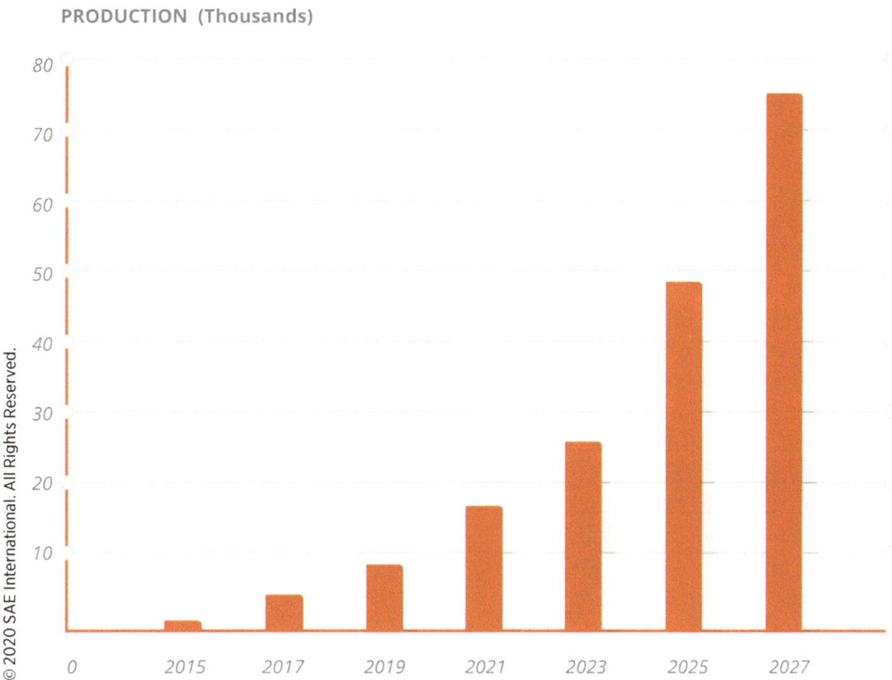

Commercial Vehicles

The commercial FCEV sector is much more mature and vibrant as compared to the passenger vehicle sector. First, significant progress in adapting the hydrogen fuel cell buses is observed and is driven by various clean city initiatives. Second, the hydrogen fuel cell trucks development is gaining momentum to meet the ambitious emission reduction targets in Europe for the heavy-duty logistics transportation sector. Finally, hydrogen material handling vehicles (MHVs) continue to grow as the most established FCEVs application.

Fuel Cell Electric Buses (FCEBs) Fuel cell electric buses (FCEBs) are gaining more attention as a potential complementary solution to battery-electric buses for future sustainable cities. Europe is leading the way with 600 FCEBs and refueling infrastructure planned within the H2BusEurope scheme. Denmark will receive 200 of these buses by 2020 [17]. The growth is expected to continue as major European cities such as London, Paris, Madrid, and Athens will only buy zero-emission vehicles starting from 2025. Besides, there are active support mechanisms in place, including €57 million (approximately $62 million) allocated to purchase 300 FCEBs and develop refueling infrastructure in 22 European cities by 2023 [13]. Major bus suppliers include Van Hool, Mercedes, Wrightbus, APTS/Phileas, VDL, and Solaris, with Ballard being the leading fuel cell stacks supplier.

In the United States, the support mechanisms for FCEBs remain strong with the Federal Transit Administration providing $90 million for the National Fuel Cell Bus Program. Around 40 vehicles are expected to be in operation by the end of 2018. California is leading the way with 25 operational buses with 10 new buses expected to join the program in 2019 [18].

In Asia, China has implemented ambitious deployment plans including a current government subsidy of ¥500,000 (or roughly $72,000) per vehicle. It is challenging to estimate the exact number of expected or in-service buses, but previous public announcements mentioned 50 FCEBs in Sichuan, and 30% of Shanghai's electric bus fleet as FCEBs. In addition, Shangdong Heavy Industry projected a production target of 2,000 buses. China also has a domestic production of FCEBs in Yunfu [19].

South Korea plans for 1000 FCEBs and a refueling infrastructure to start operation by 2022 [20] and targets all new trucks and buses to be zero-emission by 2030. The only supplier of FCEBs is Hyundai with six buses in operation and 30 additional vehicles planned for 2019. Refueling infrastructure will be provided by HyNet, through a similar funding scheme as the German H2Mobility.

Fuel Cell Electric Trucks Hydrogen fuel cells are gaining increased attention in heavy-duty mobility as they are a more attractive alternative to batteries. The main challenges with batteries is the limited driving range, considerable weight, and difficulty in fast charging a large amount of batteries. In the European Union, an additional impetus is the emission reduction targets for heavy-duty vehicles by 15% and 30% before 2025 and 2030, respectively. Nikola Motor is currently considered the global market leader in supplying fuel cell electric trucks including to a large project with Anheuser-Busch, which ordered 800 trucks to replace its existing fleet [21]. Fuel cell stacks for these vehicles are delivered by PowerCell while the powertrain and controls by Bosch. Nikola Motor is also planning to release new models for the European market (Nikola Tre) in 2022/2023 [22]. To provide refueling infrastructure, Nikola Motor has partnered with Nel Hydrogen of Oslo to build 28 large hydrogen refueling stations (8 tons of hydrogen per day) in 2020.

The European market has become pretty vibrant with Bosch announcing a partnership with stack manufacturer PowerCell in 2019 [23]. Toyota has launched an improved second-generation truck in 2018 [24] that is lighter by one ton with improved system integration. Keen interest for fuel cell trucks from the California Air Resources Board prompted it to grant funding for 10 trucks and two refueling stations to reduce emissions in ports [25].

There are also some significant developments in the midsize trucks. The flagship project involves an order for 1600 Hyundai fuel cell trucks in collaboration with H2 Energy to be delivered to Switzerland from 2019 to 2025 [26]. These trucks have an expected range of 400 km with a fuel cell system peak power output of 350 kW. They will be used by Coop, a Swiss supermarket chain. The regulatory environment in Switzerland is advantageous for hydrogen trucks as they are exempted from the roughly $70,000 annual tax imposed on a conventional 40-ton truck. Developments for medium-sized hydrogen trucks in China are progressing, but few details are available.

There is also a significant interest in fuel cell haul trucks and other heavy mining machinery. Individual major players within the mining industry believe that fuel cell haul trucks could significantly decarbonize the mining industry while enabling remote mining operations to produce fuel using local solar or wind power assets. Most haul truck manufacturers are investing in fuel cell technologies for integration into their products.

Lastly, there is a strong interest in hydrogen delivery vans from logistics companies such as FedEx and UPS, which are testing range-extender Class 6 vehicles [27]. UPS vans are equipped with 30 kW fuel cell systems from Ballard, while FedEx vehicles have fuel cell systems supplied by Plug Power. The estimated market potential is 650,000 vehicles in the United States. In Europe, Deutsche Post is planning to integrate 100 hydrogen vehicles into its fleet by 2020 [28]. In Asia, Toyota is working on trucks for 7-Eleven stores in Japan that are expected to come to the market in 2019 [29].

Material Handling Vehicles Nearly the entire market of fuel cell MHVs is concentrated in North America, with over 20,000 hydrogen fuel cell forklifts in operation [30]. Developments from other regions are gaining steam with a European forklift fleet order of more than 150 vehicles from Still by Carrefour in France [31] and an agreement between Plug Power and Linde Material Handling for fuel cell electric tow trucks and fuel cell electric forklift trucks [32]. Japan includes forklifts in their national hydrogen strategy and aims at deploying 10,000 by 2030 [33]. The current operational fleet of MHV in Japan is 100 vehicles [34]. Plug Power remains to be the leading global vehicle supplier in this sector with a few additional niche players in this market, including Hyster-Yale's Nuvera in North America, Toyota in Japan, and Fronius in Austria.

The FCEV market is proliferating and expanding to all possible segments from light vehicles to heavy mining vehicles. The growing global interest in the hydrogen economy and the improving refueling infrastructure are attractive reasons for automotive manufacturers to develop and compete with their own solutions. While larger vehicle manufacturers often have their own mature fuel cell programs and manufacture fuel cell systems in-house, other players prefer to get systems from fuel cell manufacturers and integrate them into their products.

Hydrogen Availability

Hydrogen Generation Pathways

Globally, hydrogen is a strategically important commodity, both as a primary feedstock to the refining, fertilizer, and chemical industries and as the enabler to increase the deployment of renewable energy and decarbonize the transport sector. Global hydrogen production is estimated to be between 70 and 80 million tons annually, of which 95% is extracted from fossil fuels, giving rise to more than 830 million tons of carbon dioxide emissions annually [35].

There are several other mature methods and technologies to produce hydrogen; see Figure 5. When it comes to hydrogen for FCEVs, purity, scale, and economics are key factors to ensure a reasonably priced supply of hydrogen, and therefore three main hydrogen methods are often considered:

- Steam methane reforming (SMR)
- Thermolysis of waste
- Power to gas

Other technologies, such as direct solar to hydrogen, biological or biogas reforming, and thermal processes may be viable; however, these are not presently commercially at scale or suitable for deployment at scale.

Hydrogen produced from natural gas though SMR is the most common source of bulk commercial hydrogen. In SMR, steam reacts with methane in a reformer, at high temperatures (between 700°C and 1100°C) in the presence of a metal-based catalyst (nickel-based), yielding carbon monoxide and hydrogen. This synthesis gas (syngas) exits the reformer and goes through a water-gas shift reactor that then converts the carbon monoxide to carbon dioxide and more hydrogen [37].

The final step is the separation of the hydrogen from the syngas exiting the shift reactor, which is mostly hydrogen, carbon dioxide, and water. This process is usually

FIGURE 5. **Hydrogen production pathways [36].**

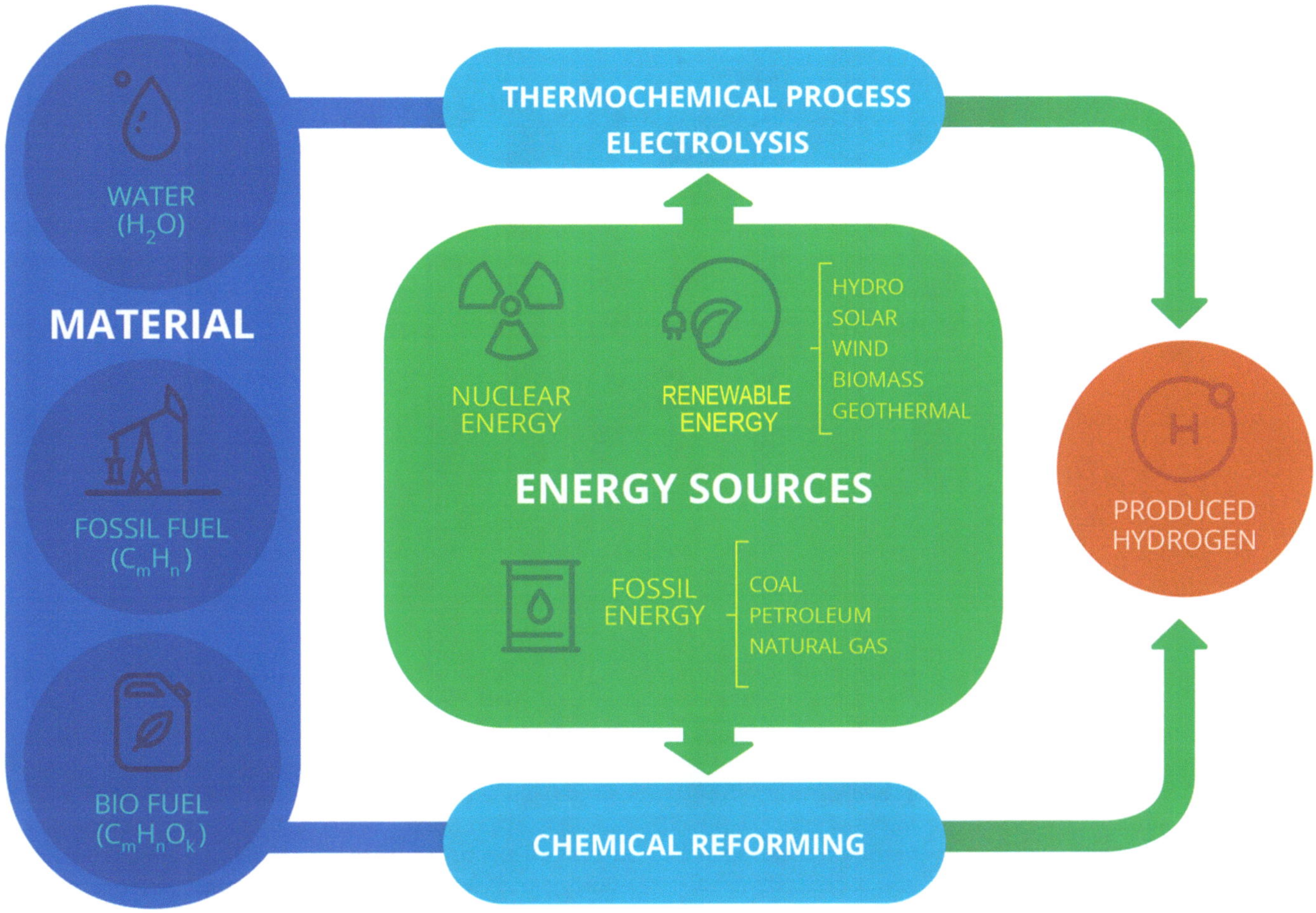

accomplished through one of the three following methods: pressure-swing-adsorption (PSA) directly separates the hydrogen, PSA then separates the carbon dioxide followed by the condensation of the remaining water, or using a membrane to separate the hydrogen, producing various degrees of end stream hydrogen purity including the fuel cell grade hydrogen (99.999% pure). Overall, the production of 1 kg of hydrogen requires approximately 2.4 kg of methane as feedstock and produces approximately 7 kg of carbon dioxide.

Thermolysis, or pyrolysis, consists of a thermal decomposition of the volatile organic components at elevated temperatures in the absence of air or oxygen. When small amounts of air are used in the process, gasification occurs at even higher temperatures that decomposes the remaining nonvolatile carbon char from pyrolysis into additional syngas [38]. Similar to the hydrogen produced from natural gas, this syngas can be suitably purified for use in FCEVs.

One of the most innovative advanced waste conversion processes available nowadays is, perhaps, the Concord Blue process developed by a U.S. company of the same name. This process is touted to be "the only commercially proven non-polluting solution that recycles waste into energy at virtually any scale." Concord Blue utilizes heat transfer instead of direct heating or combustion to convert waste into syngas and owns facilities in many locations worldwide, including Germany, India, and Japan. The facility in Omuta (Japan), which began operating in 2011, is the first biomass-to-hydrogen facility in the world, capable of converting sewage sludge into 7200 m^3 of pure hydrogen per day [39].

On the other hand, Power-to-hydrogen (or Power-to-gas) and Power-to-X technologies are very similar but with key differences. The Power-to-X technology can convert "excess," "curtailed," "constrained," or "dedicated" renewable electrical power (i.e., wind or solar) or grid electrical power to hydrogen at circa 75% efficiency using an electrolysis device at large scale (>1 MW). The X in Power-to-X technology refers to the further conversion of the produced hydrogen to other liquids, gaseous chemicals, carriers, or vectors. Therefore, a site that produces just hydrogen gas from the electrical energy can be referred to as a Power-to-gas site. If the site produces any other chemical or carrier from this hydrogen, it is Power-to-X. Power-to-gas technology has been demonstrated in over 34 projects across different European countries at multi-megawatt scales for grid

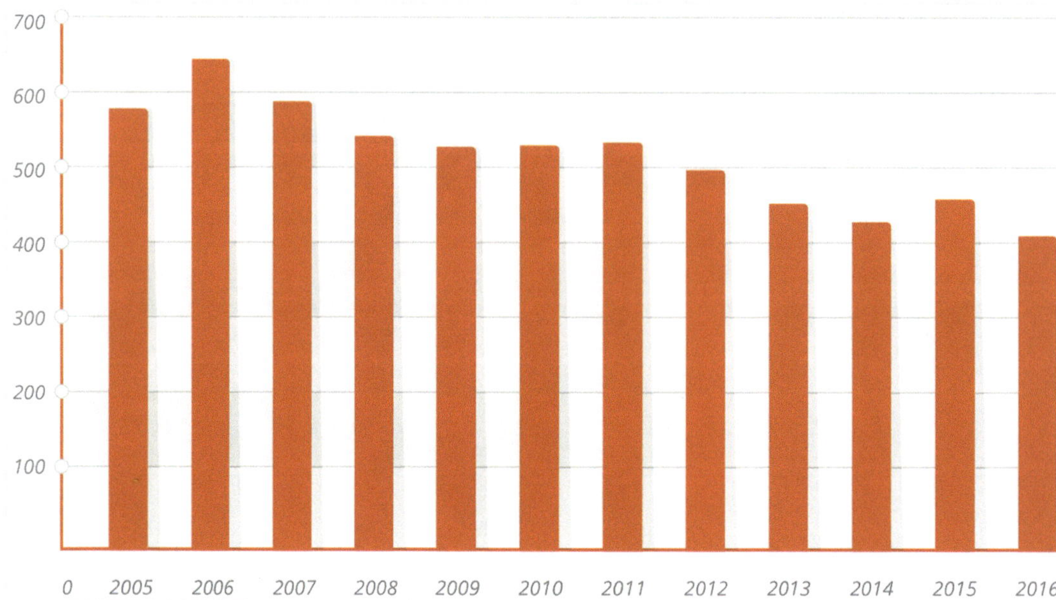

FIGURE 6. Carbon dioxide intensity of global new power generation (kg/MWh) [43].

balancing, energy storage, production of bio-methane, and production of hydrogen for industrial or transport use [40].

Power-to-gas technology is an effective way of integrating renewable electricity sources that transforms electricity into storable zero-emission and zero-carbon hydrogen by electrolysis. Placing an electrolyzer on the output of a renewable electricity source is the most basic way to produce hydrogen from renewables. However, in reality, the practicality and economics of power-to-gas technology is much more complex. There are many factors, depending on geographical or global location, site (greenfield or existing), grid connection/access cost, grid electricity cost, technology, scale, market prices of hydrogen, market prices of electricity, taxation, cost, government subsistence, and constraints (grid infrastructure, site, resources, technology, market access, capacity factor like the electrolyzer and renewable source, lifetime, others) to consider when installing an electrolysis-based hydrogen production facility with a renewable electricity source. Depending on the above physical and economic constraints, the operator of the electrolysis device may decide to convert "excess," "curtailed," "constrained," "dedicated," or "a mix" of renewable electrical power (i.e., wind or solar) to produce zero-emission and zero-carbon hydrogen.

A simpler model for hydrogen generation that is used at many refueling stations across the globe is to produce hydrogen on site not from renewable sources but directly from the electrical grid. There are distinct advantages of this model for producing a zero-emission fuel. These include

- Less transport cost of hydrogen
- Using low market price of electricity to produce hydrogen
- Storing hydrogen on site for periods of high grid prices

However, this model also suffers from various disadvantages including a non-"zero-carbon" hydrogen (as its carbon intensity is related to the carbon intensity of the grid electricity; see Figure 6) and the high cost of grid electricity compared to a renewable source site (transmission, connection, operation, and other costs). These may have a severe impact on the practicality, economics, or green credentials of using grid electricity to produce hydrogen for transport.

As a transport fuel, hydrogen produced using an average global grid electricity with a carbon intensity of 400kgCO_2eq/MWh and used in FCEVs (170 gCO_2eq/km) may have a higher carbon intensity compared to that of a BEV (70 gCO_2eq/km) and the 2021 European Union target for a fossil fuel vehicle (95 gCO_2eq/km) [41, 42].

Hydrogen Refueling Infrastructure

There are over 380 hydrogen refueling stations across (mainly) Europe, California, Japan, China, and South Korea, with the majority of hydrogen at these FCEV stations generated from fossil fuel-based processes [35].

The development of the hydrogen supply chain and refueling infrastructure is a crucial requirement for FCEVs. Depending on the choice of hydrogen production method and location compared to the hydrogen refueling station, many factors must be accounted for to ensure the supplied hydrogen is cost effective and has the correct carbon credentials.

Hydrogen refueling stations are complex multicomponent installations, and as such, they come at a very high cost. As a majority of these components are designed specifically for hydrogen applications, they are generally manufactured

FIGURE 7. **Standard on-site hydrogen generation and refueling.**

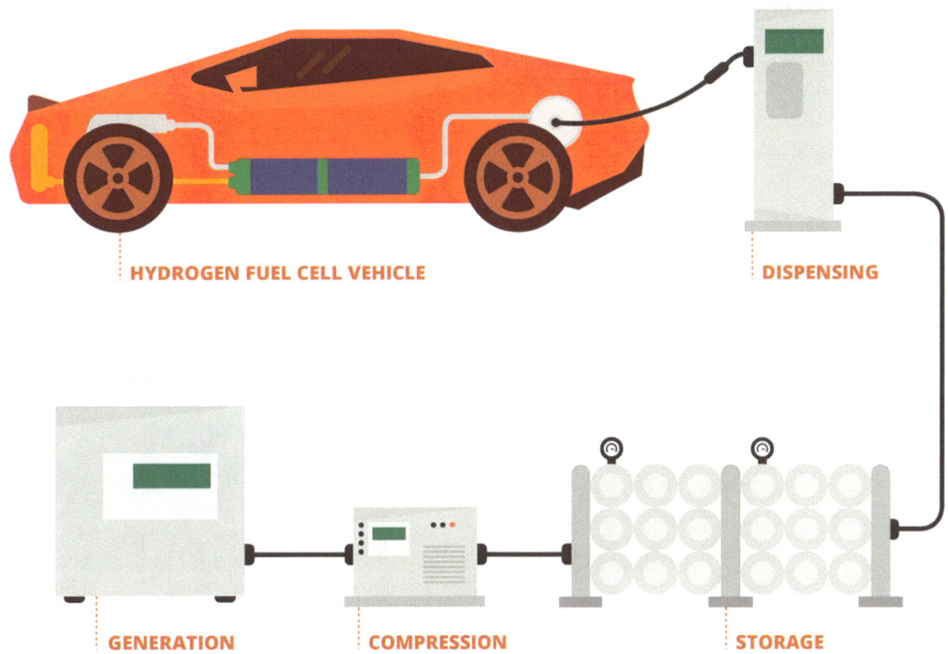

in small quantities and are very costly. Figure 7 shows the standard hydrogen refueling station with on-site hydrogen generation. Besides the electrolyzer, other components including hydrogen compressor, hydrogen storage, and hydrogen dispenser are required.

Today, a hydrogen refueling station with on-site hydrogen generation via electrolysis capable of servicing several FCEVs can easily exceed capital costs of $1 million.

Understanding Hydrogen Classification

Over the past few years, "colors" of hydrogen have entered the conversations and discussions for relating hydrogen production processes with carbon/greenhouse gas (GHG) emissions intensity. However, color schemes are inconsistent, confusing, and unclear given the many global directives at present. More clarity may be presented by the European Commission in 2020, with the RED II directive succeeding the existing regulation being implemented on January 1, 2021.

CertifHy was a European project created to develop a European Framework for the generation of guarantees of origin for green hydrogen. It has been very successful in adapting a logical approach to distinguish between and apply carbon dioxide equivalent or GHG equivalent emission values to carbon-intensive, low carbon, or zero-carbon hydrogen pathways, as shown in Figure 8.

Table 2 details some identified colors of hydrogen. Green hydrogen meets the low-carbon threshold as it uses renewable energy sources. If the GHG footprint of the generated hydrogen is below 36.4 gCO_2eq/MJ_{H2}, then 50% of it will be considered Green and would get a green label under the current guarantee of origin scheme, while the remaining 50% will be considered as a low-carbon hydrogen.

Blue hydrogen meets the low-carbon threshold but is generated using non-renewable energy sources. SMR of natural gas with carbon capture and storage (not carbon capture and utilization as the captured carbon dioxide will eventually be released to the atmosphere) as well as nuclear could be classified as blue. Currently, there is no legal value at the moment; the certification serves only to prove to the consumer the characteristics of the hydrogen. However, this European Union certification is an important first step to further develop the market for green hydrogen and provide the basis for policymakers and industry to take a leadership role in the energy transition [44].

Niche Market versus Mass Production

Decarbonization and air pollution targets are driving governments from all around the world to find innovative solutions for transportation. FCEVs, considered as complementary to BEVs, are deployed today within different transportation segments including passenger cars running long distances (i.e., Hype, hydrogen taxis in Paris), heavy-duty (i.e., Portal, hydrogen semi-trucks in California), and rail (i.e., Coradia iLint, hydrogen train in Germany) [45].

FIGURE 8. **Renewable and carbon (GHG) emission of hydrogen. Best available technology (BAT) benchmark for hydrogen generation is the SMR at 91gCO_2eq/MJ_{H2} and low-carbon threshold is set at 60% below BAT=36.4gCO_2eq/MJ_{H2} [44].**

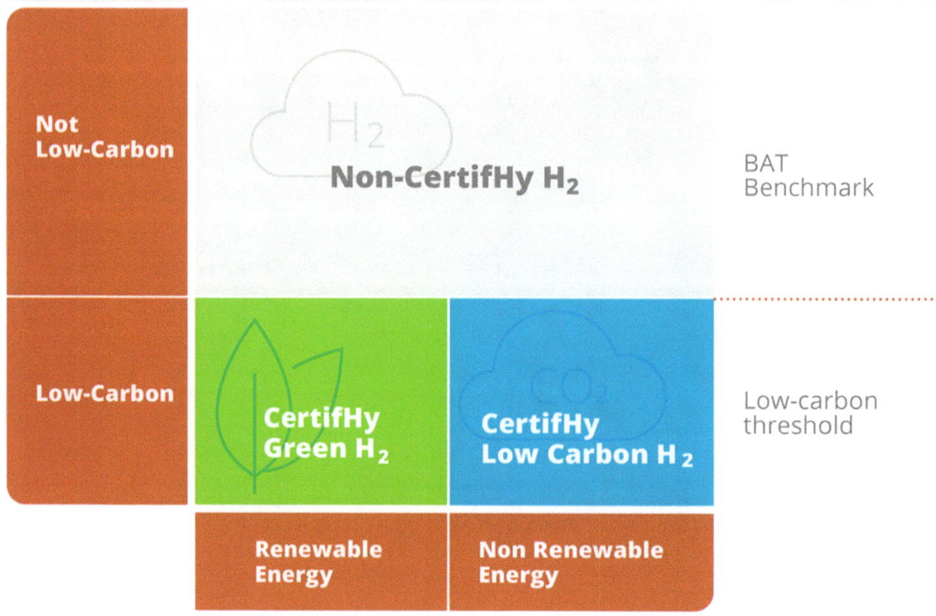

TABLE 2. **Hydrogen classification by carbon intensity color scheme.**

Hydrogen classification	Origin	Fuel type	Emissions
Black hydrogen	Fossil fuel	Coal	>91 gCO_2eq/MJ_{H2}
Brown hydrogen	Fossil fuel	Lignite	>91 gCO_2eq/MJ_{H2}
Gray hydrogen	Fossil fuel	Natural gas	91 gCO_2eq/MJ_{H2}
Blue hydrogen	Fossil fuel with carbon capture and storage	Natural gas	<36.4 gCO_2eq/MJ_{H2}
Green hydrogen	Renewable energy	-	0

Economic Drivers for Fuel Cell Vehicle Technology

Today, developing a profitable business case for hydrogen mobility is a challenge. On the one hand, there is no fair global pricing for carbon externalities, and on the other hand, carbon-based incumbent technologies benefit from the long-established optimized cost structure and operating on amortized infrastructure. Society's willingness to pay for a positive environmental externality is minimal, and the existing energy and transport system is locked into an incumbent pollutant technology.

The virtuous cycle for hydrogen technology profitability is conditional on both the cost reduction from learning-by-doing and scale effects, as well as the deployment of minimal coverage infrastructure. Once deployed locally, projects can be replicated globally to drive the associated cost curve further down and improve the global competitiveness of hydrogen technology.

Role of Learning-by-Doing and Scale Effects in Market Penetration of Fuel Cell Electric Vehicles

Today FCEVs have higher capital cost than ICEs: $60,000-$75,000 for the Toyota Mirai or Hyundai IX versus $30,000-$45,000 for a comparable Mercedes sedan, with a comparable fuel cost of $10-$12 per 100 km in both cases.

Experts believe that hydrogen and fuel cells can experience a cost reduction trajectory similar to those of solar photovoltaics and batteries, notably through learning-by-doing and scale effects. The former effect is conditioned by an increase in the cumulative volume of FCEVs produced globally and the latter, through larger installations.

Numerous studies agree that hydrogen mobility can become a cost-effective option for the reduction of carbon dioxide emissions in the long term. The horizon of cost-effectiveness differs depending on the policy support during the initial deployment stage. Without policy support, the difference in total cost of ownership (TCO) between FCEV and ICE can be brought to zero around 2040 [46], mainly due to FCEV cost reduction via learning effect and growing gasoline price. With tax incentives, FCEV could be cost-competitive with ICE as early as 2025 [47].

Demand and Subsidies as a Major Market Disruptions

The critical number of stations and vehicle stock to kick-start the FCEV market could be attained through two mechanisms: first through a vehicle deployment push and second by an infrastructure deployment pull. Indeed, the governments could implement policy instruments targeting either vehicles or infrastructure deployment to solve the chicken-and-egg dilemma [48].

In the first case, the government acts mainly on the demand side of the problem and encourages consumers by creating monetary and non-fiscal incentives to make FCEV possession more attractive. The associated demand for refueling is supposed to push the corresponding infrastructure deployment. In the second case, the government tackles the supply side of the problem by inducing infrastructure deployment through subsidies or coordinated public-private partnerships which, in turn, will increase the demand for FCEVs.

In many actual cases, governmental support for both vehicle supply and infrastructure deployment is observed at the same time in countries like Denmark and Japan.

Availability of Hydrogen Refueling Infrastructure

Two strategies to overcome the chicken-and-egg dilemma have been used for FCEVs. The first one is to build a captive fleet. A captive fleet is a group of vehicles, such as a government or company fleet of delivery vehicles used for a dedicated purpose. This approach greatly facilitates the forecasting of fuel consumption and the deployment needs for the refueling network. Significant early markets are created by specialized application niches, such as the Parisian Hype taxi fleet, and by the political wills of municipal early adopters. The second strategy relies on public subsidies to quickly set up an extensive infrastructure, with a possible initial focus on clusters before expanding to interregional roads, as executed within H_2 Mobility program in Germany. In this case, the government expects to gain the trust of car manufacturers and customers and reduces the uncertainty related to the FCEV deployment.

Today, major OECD countries are driving the hydrogen infrastructure deployment effort: there are 34 hydrogen stations in California, 100 in Japan, 12 in South Korea, 50 in Germany, 6 in France, and 12 in Denmark.

Social Aspects of Commercial Hydrogen Rollout and Hydrogen Education

Breaking Social Perception of Hydrogen

Social perception of harnessing hydrogen as a fuel is generally very diverse. On one hand, there are proponents of this technology who believe that the technology is safe as long as it is used in accordance with rules, regulations, and safety standards explicitly developed for hydrogen technologies. On the other hand, there are hydrogen opponents who believe that dealing with highly flammable and compressed gas is a hazard no matter what safety standards are in place. While hydrogen has to be handled with care, the next section of this report compares properties of hydrogen with natural gas (Table 3). There is an ongoing debate among experts and

TABLE 3. Comparison of flammability and autoignition temperatures for selected fuels.

	Lower flammability level in percentage by volume of air	Upper flammability level in percentage by volume of air	Autoignition temperature
Hydrogen	4%	75%	550°C
Diesel	0.6%	7.5%	210°C
Gasoline	1.4%	7.6%	260°C
Propane	2.1%	9.5%	480°C
Ethanol	3.3%	19%	365°C
Methane	5%	15%	580°C
Kerosene (Jet A-1)	0.7%	5%	210°C

FIGURE 9. **Public perception of fuel and its automotive applications. Reproduced from the World Economic Forum [52].**

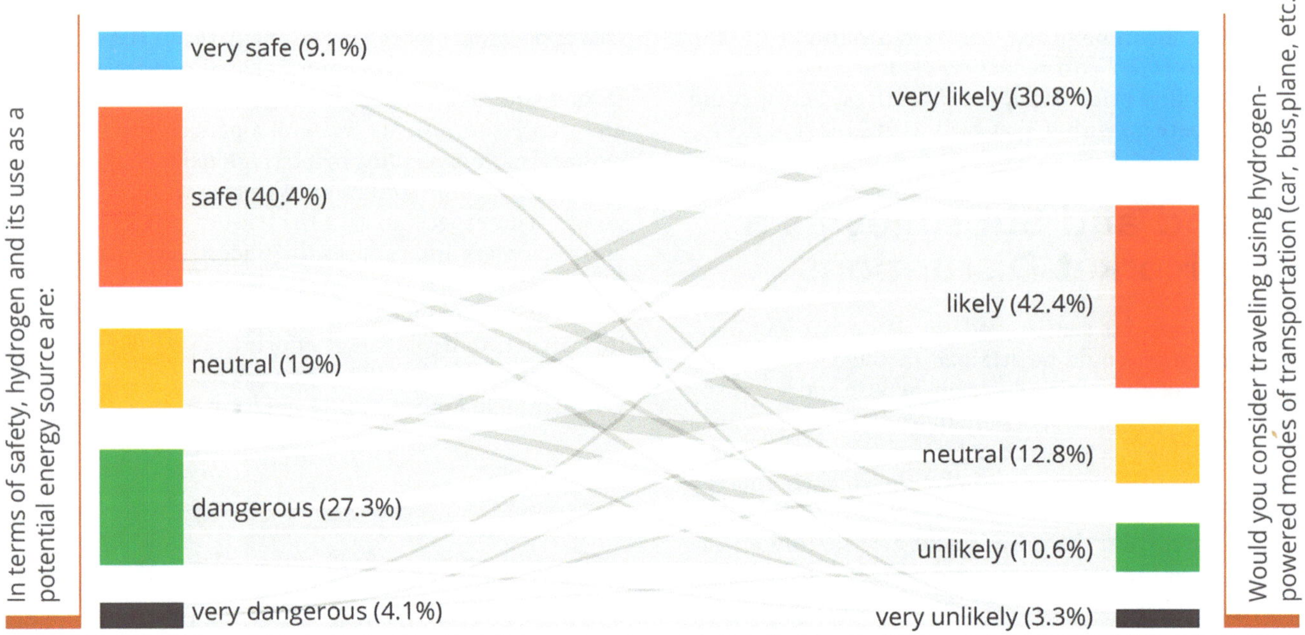

technology proponents whether hydrogen is safer or more dangerous than traditional fossil fuels like petrol or diesel. However, the social perception of hydrogen, just like many other emerging technologies, is often influenced by incidents during early implementation stages.

A clear example of such accidents that negatively impacted hydrogen on a long term was the infamous Hindenburg disaster. The Hindenburg disaster occurred in 1937 and resulted in 37 fatalities. Certain studies, however, attributed the Hindenburg disaster not to hydrogen itself but to the formation of electrical discharge caused by built-up static electricity in the airship, which ignited the outer skin of the Hindenburg. This outer coating layer on the Zeppelin was made from a mixture of highly flammable iron oxide and aluminum paint [49]. To this day, the Hindenburg disaster is still used as an argument against when discussing hydrogen mobility applications.

In Australia, especially in the Australian state of Queensland, the slow adaptation of hydrogen in automotive applications is often attributed to a failed vehicle which was heralded by its inventor to generate hydrogen from water and burn this hydrogen on board in a controlled nuclear fusion process to power the vehicle. This mysterious invention made national headlines when it was publically showcased in Brisbane in 1980 before the general public and journalists as well as former Queensland Premier Sir Joh Bjelke-Petersen [50]. Today, after nearly 40 years, this failed hydrogen car is often referred to by the public when talking about hydrogen.

The recent wave of hydrogen developments and implementation projects coincided with the global headlines from South Korea, regarding that country's development and testing of a hydrogen bomb [51]. The 2017 revelation was quite unfortunate for the emerging hydrogen industry since some skeptics wrongly associated hydrogen with weapons of mass destruction. Thermonuclear weapons are often colloquially called hydrogen bombs, or H-bombs. However, they have little to do with the hydrogen used in the automotive or other industries. Thermonuclear weapons consist of a uranium- or plutonium-fueled nuclear fission primary stage and a separate nuclear fusion secondary stage fueled by heavy hydrogen isotopes (deuterium and radioactive tritium).

The above, among other unfortunate or often wrongly perceived, hydrogen applications and accidents are often deeply rooted in the minority of the society. This perception often prevents various hydrogen developments. Luckily, the majority of the population is supportive of the hydrogen economy (Figure 9), especially the idea of applying clean fuel to fight climate change. Breaking the social perception of hydrogen hazards is a difficult task which requires continuous engagement and education.

Developing Technological Understanding, Skills, and Maintenance Know-How

Social perception of hydrogen is closely interlinked with education and the understanding of the technology and its limitations. As the market share of hydrogen and fuel cell technologies is still relatively small, it is expected that the general public is unfamiliar with these concepts. For example, it often comes with great difficulty to explain to the public that fuel cells don't burn hydrogen like petrol in the ICEs, but convert hydrogen and oxygen in an electrochemical process to directly generate electricity which is then used in the electric motor to power the vehicle. This leads to another technological misconception, where a significant percentage of the society is convinced that the only electric vehicles are those powered by batteries. Education and public engagement have a vital role to play to build understanding and change existing misconceptions. These misconceptions will likely fade away and technology understanding will grow naturally as hydrogen and fuel cell applications gain more significant market penetration.

The emerging hydrogen economy will require specialized, often highly educated personnel to design, operate, and maintain hydrogen infrastructure, and also to maintain and repair automotive fleets. As the hydrogen economy emerges, educational institutions should revisit their curriculum to meet the demand of this future industry. Additionally, a better marketing effort needs to be performed by hydrogen companies to convince the public.

Hydrogen Handling and Safety Aspects

Hydrogen safety concerns cover issues related to safe hydrogen production, handling, storage, and use. The safety issues related to hydrogen differ depending on the state of hydrogen. Gaseous hydrogen is generally compressed to high or very high pressures to improve the energy density of the fuel. The high pressure of compressed hydrogen raises safety concerns by itself. Liquid hydrogen poses additional challenges due to its increased density and the extremely low temperatures needed to keep it in liquid form. Liquefied hydrogen allows for further improvement of energy density but it must be cooled below the hydrogen's critical point of 33 K (-240.15°C) [53]. Due to its ultralow temperature, liquid hydrogen is a hazard for cold burns, just like the more commonly used liquefied nitrogen counterpart. While hydrogen is biologically inert, rapidly heating and vaporization of liquefied hydrogen may cause oxygen displacement leading to oxygen deprivation and asphyxiation, which can result in fatality. Whether liquid, compressed, or at atmospheric pressure, hydrogen is flammable (Figure 10).

While attempts have been made in the past to use liquefied hydrogen as a fuel, liquefaction and the need to keep liquefied hydrogen at very low temperatures poses a significant engineering and storage challenge. Liquid hydrogen shows a promise for use in space applications because of energy density per unit mass and energy density per unit volume. Hydrogen atoms are the smallest elemental atoms in the universe, allowing molecular hydrogen to slowly penetrate and leak through any material and containment vessels. SAE J2578 covers issues related to hydrogen leakage in more detail [54]. Because of this unusual property of hydrogen, significant efforts have been made to store hydrogen in chemical form, and liberate it on demand. Varieties of chemicals, including ammonia, metal hydrides, toluene, etc., have been explored. These chemicals form covalent bonds with

FIGURE 10. Potential flammability of hydrogen discharges in air. Reproduced from SAE International [54].

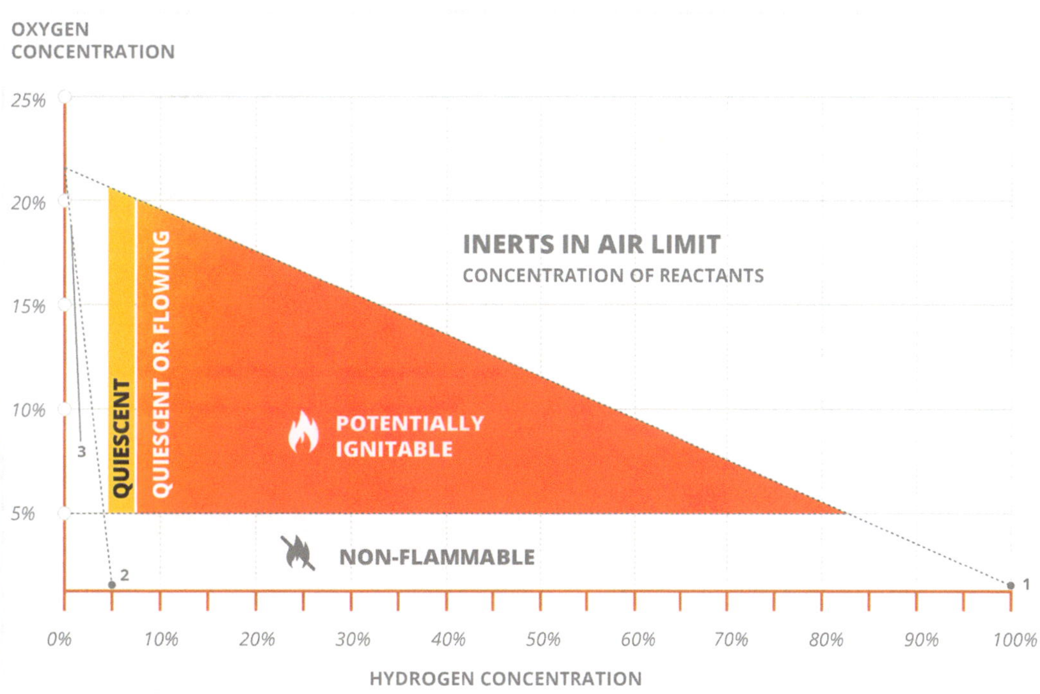

FIGURE 11. **Density of hydrogen in different states. Reproduced from BloombergNEF [55].**

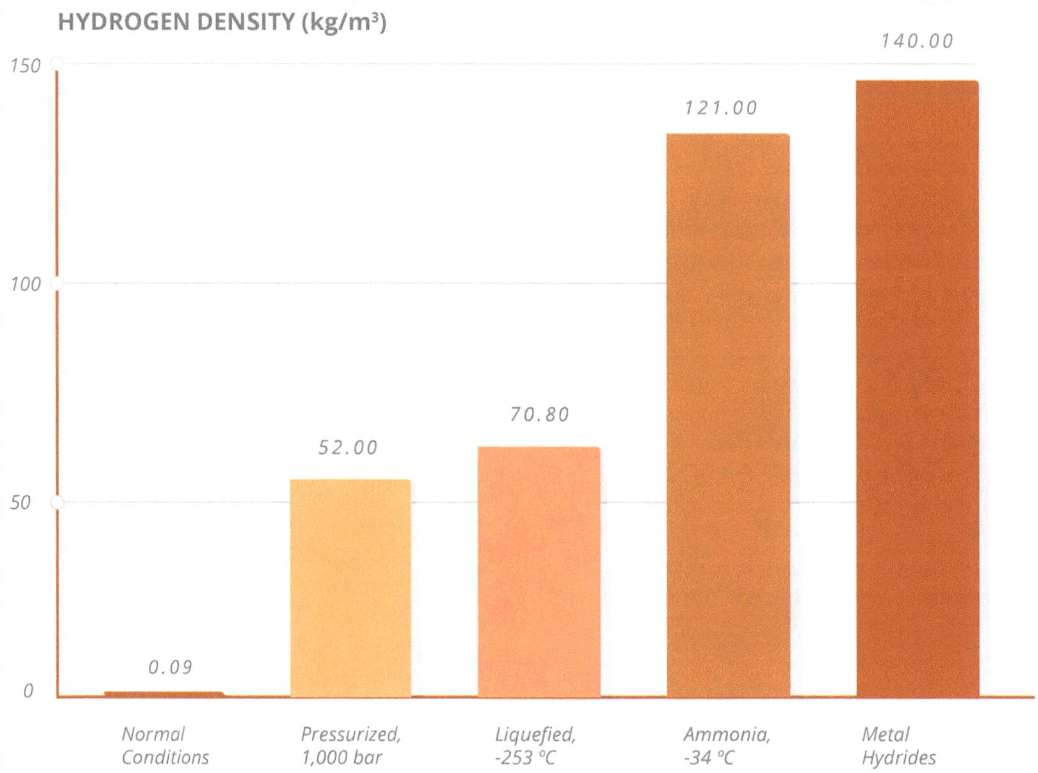

hydrogen, allowing for safe storage of hydrogen in liquid or even solid state. However, none of these solutions are commercially applied mainly because of complex processes required for hydrogen fixation and on-demand liberation, as well as poor cyclability of these storage compounds [55]. Commercial fuel cell vehicles rely on compressed hydrogen fuel, as such the remainder of this section will focus on the safety aspects of compressed hydrogen (Figure 11).

There are two broadly accepted and applied compression standards for hydrogen, 350 bar (22 kg/m^3) and 700 bar (38 kg/m^3).[56] Both of these compression standards are used in automotive applications. Generally, fuel cell light vehicles utilize hydrogen fuel compressed to 700 bar. This is purely because of the space limitation in light vehicles, where the 700 bar can store nearly twice much energy in the form of hydrogen than the 350 bar. Its drawbacks are more costly storage tanks and safety issues in storing high pressure flammable gas. Furthermore, the hydrogen at dispensing stations is stored at even higher pressures than those found in vehicles. For 350 bar vehicles, hydrogen at the station is compressed to roughly 450 bar; whereas for 700 bar vehicles hydrogen at the station is compressed to 900 bar or more (Figure 12).

Regardless, hydrogen has properties that are safer than many of the derived fossil fuels currently used in ICE vehicles. The majority of fossil fuels come in liquid form, which is potentially dangerous when such fuels spill and catch on fire. Hydrogen is flammable, but as the lightest element in the universe, hydrogen quickly ascends and dissipates into the atmosphere. When burning, conventional carbon-based fuels create radiant heat and produce hot ash. Burning of hydrogen produces little radiant heat and no ash, just pure water vapor.

The Role of Regulations and Safety Standards

As a highly compressed and flammable gas, hydrogen requires clear and stringent rules in relation to hydrogen generation, handling (including refueling), storage, and use. Effectively, hydrogen safety issues come down to three major aspects:

- Safety related to flammable gases
- Safety related to highly compressed gases
- Hydrogen leakage

SAE International and other standardization bodies globally have been working to ensure best practices and standards in terms of hydrogen generation and usage, as well as to advise on the adequacy and suitability of materials for hydrogen storage solutions. While new materials and methodologies for hydrogen storage are areas of continuous development and interest, hydrogen leakage through contained vessels is still to be addressed. It is important to

FIGURE 12. Hydrogen density at different temperatures and pressures. Reproduced from BloombergNEF [55].

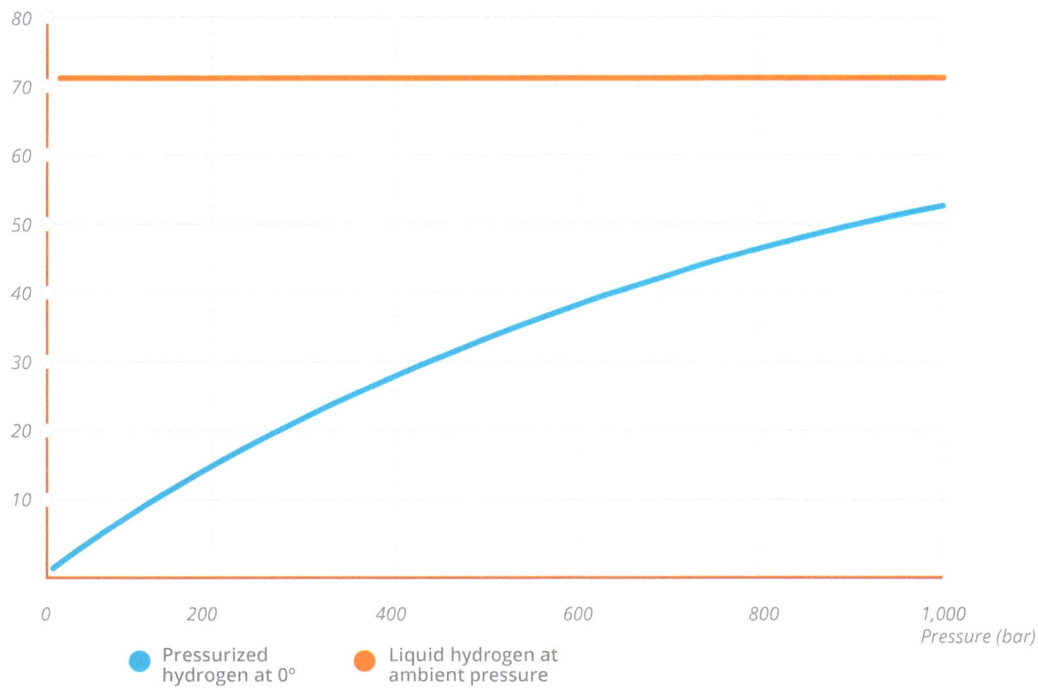

note that hydrogen cannot be stored in all types of materials, i.e., hydrogen is known to generate embrittlement and material failure in certain types of steel materials. As such, it is crucial to have clear regulations and standards to ensure safe hydrogen applications.

Another area related to hydrogen safety that is yet to be addressed is having fast and precise hydrogen sensing. Today, all hydrogen systems should be equipped with hydrogen and fire sensors to notify the control system of potential fire hazards coming from a hydrogen leakage to shut down the system. It is generally accepted that hydrogen sensors should be able to detect hydrogen concentration at 50% of the lower flammability level. However, there are certain challenges in detecting the smallest molecule in the universe. Especially for mobile applications detecting hydrogen leakage in moving vehicle proves to be a significant challenge.

The U.S. Department of Energy sets and regularly updates safety milestones for hydrogen technologies. OEMs are continually improving the technology and enhancing safety standards. It is essential that global standardization agencies take part in this process to ensure the best possible up-to-date practices. It is crucial that these standards evolve as the technology matures. It is also important that the standards are globally accepted and recognized to ensure technology transferability between national borders.

SAE International has done significant work to ensure safe and up-to-date practices that include matters related to hydrogen safety, hydrogen refueling standards, and many other aspects on using hydrogen. These are presented in Table 4.

Report Conclusions and Recommendations

Need for Partnerships and Consolidated Implementation Approaches

There are already many commercially available fuel cell light vehicle and fuel cell bus solutions - a number of other fuel cell-based ground vehicle applications are currently being developed, these include four-wheel-drive vehicles, light trucks, semi-trucks, and heavy trucks, as well as fuel cell trams, trains, and mining vehicles. Arguably, the automotive industry has never been more convinced that hydrogen is a feasible solution. This unprecedented transformation to clean energy solutions in the automotive industry has, however, its unique challenges. Predominantly, the transition to hydrogen-based transportation requires extensive engagement and partnerships by stakeholders from the entire hydrogen value chain. Equipment manufacturers and infrastructure developers must work closely with fuel distributors (B2C), as well as vehicle manufacturers to put projects together and ensure vehicle and fuel availability and also sufficient hydrogen refueling infrastructure. It is generally believed that the demand for hydrogen in a number of countries will be initially driven by corporate customers and city councils. Corporate customers,

TABLE 4. SAE publications related to hydrogen and fuel cells in the ground vehicle applications.

Publication	Title	Description
SAE J2574	Fuel Cell Vehicle Terminology	Provides definitions for hydrogen fuel cell-powered vehicle terminology. The document is intended to be a resource for other hydrogen fuel cell vehicle documents, specifically, standards or recommended practices.
SAE J2760	Pressure Terminology Used in Fuel Cells and Other Hydrogen Vehicle Applications	Provides recommended practices for fuel systems in fuel cell and other hydrogen vehicles. As part of this work, definitions for pressurized systems and containers were developed.
SAE J2601	Fueling Protocols for Light Duty Gaseous Hydrogen Surface Vehicles	Establishes the protocol and process limits for hydrogen fueling of light-duty vehicles. These process limits (including the fuel delivery temperature, the maximum fuel flow rate, the rate of pressure increase, and the ending pressure) are affected by factors such as ambient temperature, fuel delivery temperature, and initial pressure in the vehicle's compressed hydrogen storage system.
SAE J2601/2	Fueling Protocol for Gaseous Hydrogen Powered Heavy Duty Vehicles	Provides performance requirements for hydrogen dispensing systems used for fueling 35 MPa heavy-duty hydrogen transit buses and vehicles, with other pressures being optional.
SAE J2601/3	Fueling Protocol for Gaseous Hydrogen Powered Industrial Trucks	This document establishes safety limits and performance requirements for gaseous hydrogen fuel dispensers used to fuel hydrogen-powered industrial trucks. Besides, the document provides examples of fueling methods for gaseous hydrogen dispensers serving hydrogen-powered industrial truck vehicles.
SAE J2579	Standard for Fuel Systems in Fuel Cell and Other Hydrogen Vehicles	Defines design, construction, operational, and maintenance requirements for hydrogen fuel storage and handling systems in on-road vehicles.
SAE J2719	Hydrogen Fuel Quality for Fuel Cell Vehicles	Provides background information and a hydrogen fuel quality standard for commercial PEM fuel cell vehicles.
SAE J2615	Testing Performance of Fuel Cell Systems for Automotive Applications	Provides a framework for performance testing of fuel cell systems designed for automotive applications with direct current output.
SAE J2572	Recommended Practice for Measuring Fuel Consumption and Range of Fuel Cell and Hybrid Fuel Cell Vehicles Fueled by Compressed Gaseous Hydrogen	Establishes uniform procedures for testing fuel cell and hybrid FCEVs, excluding low-speed vehicles, designed primarily for operation on the public streets, roads, and highways. The procedure addresses those vehicles under test using compressed hydrogen gas supplied by an off-board source or stored and supplied as a compressed gas onboard.
SAE J2578	Recommended Practice for General Fuel Cell Vehicle Safety	Identifies and defines requirements relating to the safe integration of the fuel cell system, the hydrogen fuel storage and handling systems and high voltage electrical systems into the overall fuel cell vehicle. The document may also be applied to hydrogen vehicles with ICEs.
SAE J2616	Testing Performance of the Fuel Processor Subsystem of an Automotive Fuel Cell System	Provides design verification procedure. In addition, the documents may be used to verify design specifications or vendor claims. Test procedures, methods, and definitions for the performance of the fuel processor subsystem of a fuel cell system are provided.
SAE J3089	Characterization of On-Board -Vehicular Hydrogen Sensors	Provides test methods for evaluating hydrogen sensors when the hydrogen system integrator and vehicle manufacturer elect to use such devices on board their hydrogen vehicles, including hydrogen FCEVs.
ISO 23273	Fuel Cell Road Vehicles - Safety Specifications - Protection against Hydrogen Hazards for Vehicles Fuelled with Compressed Hydrogen	Specifies the essential requirements for fuel cell vehicles concerning the protection of persons and the environment inside and outside the vehicle against hydrogen-related hazards.
ISO/DIS 14687	Hydrogen Fuel Quality - Product Specification	Specifies the minimum quality characteristics of hydrogen fuel for utilization in automotive and stationary applications.
ISO 17268	Gaseous Hydrogen Land Vehicle Refueling Connection Devices	Defines the design, safety, and operation characteristics of gaseous hydrogen land vehicle refueling connectors.
ISO/TS 19880- 1	Gaseous Hydrogen - Fuelling Stations - Part 1: General Requirements	Recommends the minimum design characteristics for the safety and performance of gaseous hydrogen refueling stations for light-duty land vehicles.

often, service large enough vehicle fleets to make a viable business case in investing in hydrogen refueling infrastructure. Similarly, local governments or public transport providers often service bus fleets of size significant enough to make a business case for investing in hydrogen refueling infrastructure. This initial infrastructure is, often, open to external customers, including private customers who drive the demand for further hydrogen refueling points extending the infrastructure coverage.

Governments also play an important role in this stakeholder mix by offering subsidies, tax benefits, and other financial support for infrastructure development and infrastructure operational expenditure. While it is not yet a common practice, certain governments also provide significant subsidies for purchasing FCEVs, effectively making them more attractive and cost-competitive.

The so-called "chicken or egg" paradox is often referred to when discussing the implementation of hydrogen mobility projects. Contributors to this report agree that hydrogen infrastructure implementation should come first. Once hydrogen source is available locally, interest in vehicle procurement will follow. Therefore, partnerships and consolidated implementation approaches are essential in hydrogen mobility projects. Before hydrogen infrastructure is developed, it should be established whether there is a demand for it. Big anchor tenants, such as bus or truck or car fleets, are a valuable contributor to hydrogen economy, making the cost of infrastructure worthwhile.

Uniform and global technology standards also fall under the consolidated implementation approach. International standards must be in place to ensure universal technological compatibility, i.e., universal refueling standards for FCEVs. SAE International is at the forefront to establish global standards to enable this transition to clean and safe hydrogen transportation. SAE International is working with industry leaders to identify and establish the best, most efficient, and safe practices in the FCEVs space.

Over the years, SAE International has developed standards and practices intended to address the growing interest and demand in hydrogen vehicles and to guide and serve the industry towards safe and effective hydrogen refueling, storage, handling, and uses (SAE's Standards Development Program). These practices are listed in Table 4.

SAE International has formed a working group, i.e., Hydrogen Quality Task Force (HQTF) and Fuel Cell Standards Committee, specifically tasked with the standardization of hydrogen and fuel cell technologies in automotive applications. The goal of these two working groups is to

- Avoid competing documents and studies
- Eliminate costly duplication of effort
- Streamline and accelerate the development process
- Create harmonized taxonomies and standards

Next Steps for Fuel Cell Electric Ground Vehicles

Hydrogen fuel cell vehicles continue to gain more recognition as part of the sustainable future transportation. Although still at its infancy stage, the market is very vibrant across all sectors. Nearly all major OEMs have announced plans or already released their first FCEV models. The most substantial growth comes currently from the commercial vehicle sector including buses, trucks, and MHVs, while the passenger vehicles face competition from the battery electric solutions. The United States is presently the market leader in hydrogen passenger vehicles, trucks, and MHVs, while Europe has a strong position in the buses segment driven by environmental targets. China could be a giant potential market but has a lot of uncertainty on the current hydrogen FCEV status.

Due to the relative immaturity of the field, there are still a number of issues to be addressed, most of which have been briefly discussed in this publication. This publication should be considered as an introduction to the different topics identified as issues in the FCEVs. The SAE EDGE™ Research Report series intends to open a discussion among stakeholders to address outstanding technical and safety concerns. It is expected that these initial discussions will lead to the formation of working groups or standardization committees with an ultimate goal of developing frameworks, regulations, ontologies, best practices, and standards to enable continuous progress toward safer and more innovative automotive products and solutions.

SAE International has an extensive track record in developing automotive standards in various industry segments with some pioneering work related to the hydrogen and fuel cell applications in the automotive industry. Basic definitions and terminology (SAE J2574), frameworks covering hydrogen safety (SAE J2578), and protocols for hydrogen refueling (SAE J2601, SAE J2601/2, and SAE J2601/3), among many others, are broadly applied, referenced, and considered globally as best practices.

While SAE International has done extensive work (some listed in Table 4) to standardize the field of hydrogen and fuel cell applications in ground vehicles, the evolving nature of this area requires continuous development to keep up with changing industry trends and technological innovations. This SAE EDGE™ Research Report on unsettled issues concerning the use of fuel cells in electric ground vehicles identified the following critical issues for further discussions among industry stakeholders, topical working groups, and dedicated standardization committees:

- The need for comprehensive and standardized ontologies and vocabularies with precise definitions in the area of hydrogen and fuel cells
- The need for a standardized set of protocols and methodologies for ensuring safe hydrogen

generation, storage, refueling, and utilization in FCEVs

- The need for clearly defined and standardized fuel cell systems and components, including hydrogen storage systems; for automotive applications, in terms of system size, power density, conversion efficiency, etc.
- The need for clearly defined standards and protocols for hydrogen sensing, especially in the context of vehicle applications covering parameters such as sensor response time, selectivity, lower detection limits, etc.
- The need for standardized classification of hydrogen origin and hydrogen certification standards

Recommendations

The overall recommendations of this SAE EDGE™ report can be summarized as follows:

1. Consolidated approaches and effective partnerships, including stakeholders along the entire hydrogen value chain, are necessary and essential to enable the transition into hydrogen-based transportation.
2. Common standards and practices for hydrogen, fuel cells, hydrogen refueling methodologies, and refueling infrastructure are highly desirable to allow technological universality and broad technology intake.
3. Evolving safety standards for hydrogen generation, compression, storage, handling, and refueling should be continuously evaluated and improved to ensure the operational safety of the vehicles and refueling infrastructure.
4. Educational activities around technological capabilities and hydrogen safety should play a major role in changing the societal perception of hydrogen to enable broad and rapid technology intake.
5. Governmental support in the form of subsidies, incentives, carbon tax, or other forms of financial aid is essential at this point to allow hydrogen and fuel cell transportation to be cost-competitive with traditional fossil fuel-based economy.
6. Further policy developments, including the development of standards, should leverage existing capabilities of various international, governmental, not-for-profit, and industry partners.

The overarching goal of the above recommendations is to enable rapid technology intake while ensuring technological safety.

SAE EDGE™ Research Reports

SAE EDGE™ Research Reports, like the present report on "Unsettled Issues Concerning the Use of Fuel Cells in Electric Ground Vehicles," are intended to push further out into still unsettled areas of technology of interest to the mobility industry. SAE launches these reports before attempting to form a joint working group, let alone a Cooperative Research Program or a Standards Committee.

SAE EDGE™ Research Reports are intended to be quick, concise overviews of major unsettled areas where vital new technologies are emerging. By definition, an unsettled area is characterized more by confusion and controversy than established order. Early practitioners must confront an absence of agreement. Their challenge is often not to seize the high ground but to find common ground. These scouting reports from the frontiers of investigation are intended merely to begin the process of sorting through critical issues, contributing to a better understanding of key problems, and providing helpful suggestions about possible next steps and avenues of investigation.

SAE EDGE™ Research Reports, therefore, are fundamentally distinct from the more formal working groups approach and far removed from the more mature Cooperative Research Program and standard's development process. They begin by convening an informal meeting of subject matter experts working in an unsettled area of technology from the industry, government, and the research community. A moderated but collegial discussion ensues regarding their shared interest. This conversation attempts to identify critical issues hindering a clear understanding of important topics, preventing a common agreement on potential solutions, or slowing progress in key areas of research.

An SAE EDGE™ Research Report is then rafted and circulated for comment to meeting attendees and other interested parties. After review, the final result of this process is a report like the current examination of unsettled issues concerning the sensors needed for automated vehicles.

Abbreviations/Definitions

S - Aktieselskab. A Denmark stock-based company

B2C - Business-to-consumer. Process of selling products and services directly between consumers or end-users

BAT - Best available technology. Technology approved by regulators or legislators for meeting output standards for a particular process

BEV - Battery electric vehicle

BMW - Bayerische Motoren Werke. German multinational automotive company

CO_2 - Carbon dioxide. Gaseous chemical compound at room temperature and pressure. Major climate change contributor

FCEB - Fuel cell electric bus

FCEV - Fuel cell electric vehicle

FedEx - FedEx Corporation. American multinational courier delivery services company

G20 - Group of Twenty. An international forum for the governments and central bank governors from 19 countries and the European Union

gCO_2eq/km - Grams of carbon dioxide equivalent per kilometer. Measure of carbon dioxide released per traveled distance

gCO_2eq/MJ_{H2} - Grams of carbon dioxide equivalent per megajoule of hydrogen energy. A measure of carbon dioxide released per energy in the form of hydrogen. This measure relates to conversion technologies which utilize fossil fuels to generate hydrogen

GHG - Greenhouse gas

H_2 - Chemical formula for hydrogen

hp - Horsepower. Imperial unit of power.

HQTF - Hydrogen Quality Task Force. Task force of the Interface Working Group of the Fuel Cell Standards Committee at the Society of Automotive Engineers

ICE - Internal combustion engine

K - Kelvin. Unit of temperature

kg - Kilogram. Unit of weight

kg/m^3 - Kilogram per cubic meter. A measure of density

kg/MWh - kilogram per megawatt-hour. A measure of the weight of produced goods per electrical power consumption over time

$kgCO_2eq/MWh$ - Kilograms of carbon dioxide equivalent per megawatt-hour. A measure of carbon dioxide emissions per unit of generated energy

km/h - Kilometers per hour. A measure of speed

LMPH2G - Prototype fuel cell-powered race car developed by Swiss automotive company GreenGT

MHV - Material handling vehicle

MIT - Massachusetts Institute of Technology

MRL - Manufacturing readiness level

MW - Megawatt. Unit of power

NASA - National Aeronautics and Space Administration. Independent agency of the United States Federal Government responsible for the civilian space program, as well as aeronautics and aerospace research

NEESC - Northeast Electrochemical Energy Storage Cluster

NEL - NEL ASA is a Norwegian company providing solutions for the production, storage, and distribution of hydrogen

OECD - Organisation for Economic Co-operation and Development. An intergovernmental economic organization founded to stimulate economic progress and world trade

OEM - Original equipment manufacturer

PEM - Proton exchange membrane. A fuel cell type

PHEV - Plug-in hybrid electric vehicle

PSA - Pressure swing adsorption. The process used to separate gas species from a mixture of gases under pressure according to species' molecular characteristics and its affinity for an adsorbent material

Pte Ltd - Private limited company. In Singapore, a private company with a limited guarantee

RED II - Renewable Energy Directive II. Second iteration of the European Union directive which mandates levels of renewable energy use within the European Union

SMR - Steam methane reforming. A chemical process in which methane is converted into syngas (hydrogen and carbon monoxide)

TCO - Total cost of ownership. Financial estimate intended to help buyers determine the direct and indirect costs of the product

TRL - Technology readiness level

U.S. - United States [of America]

UPS - United Parcel Services. American multinational package delivery and supply chain management company

Works Cited

1. Salameh, M.G., "How Viable Is the Hydrogen Economy? The Case of Iceland," International Association for Energy Economics, 2008, https://www.iaee.org/en/publications/newsletterdl.aspx?id=59.
2. Warshay, M. and Prokopius, P.R., "The Fuel Cell in Space: Yesterday, Today and Tomorrow," NASA, 1989, https://ntrs.nasa.gov/archive/nasa/casi.ntrs.nasa.gov/19900002488.pdf.
3. Room, J.J., "The Hype about Hydrogen," *Issues in Science and Technology* 20(3):74-81. https://www.jstor.org/stable/43312457.
4. Wald, M.L., "U.S. Drops Research Into Fuel Cells for Cars," *The New York Times*, 2009. https://www.nytimes.com/2009/05/08/science/earth/08energy.html.
5. Paris Agreement, "United Nations Treaty Collection," July 8, 2016, https://treaties.un.org/pages/ViewDetails.aspx?src=TREATY&mtdsg_no=XXVII-7-d&chapter=27&clang=_en (Retrieved July 28, 2019).
6. Thomas, S., "Fuel Cell and Battery Electric Vehicles Compared," 2009, https://www.energy.gov/sites/prod/files/2014/03/f9/thomas_fcev_vs_battery_evs.pdf.

7. U.S. Department of Energy, "FY 2012 Annual Progress Report: IV.E.5 Early Market TRL/MRL Analysis," DOE Hydrogen and Fuel Cells Program, 2012, https://www.hydrogen.energy.gov/pdfs/progress12/iv_e_5_ronnebro_2012.pdf.
8. U.S. Department of Energy, "Fuel Cell Technologies Market Report," Fuel Cell Technologies Office, 2017, https://energy.gov/sites/prod/files/2017/10/f37/fcto_2016_market_report.pdf.
9. "Global Market for Hydrogen Fuel Cell Vehicles," 2018, https://www.prnewswire.com/news-releases/close-to-6500-hydrogen-fuel-cell-vehicles-have-been-sold-globally-300594992.html.
10. California Environmental Protection Agency- Air Resources Board, "Annual Evaluation of Fuel Cell Electric Vehicle Deployment and Hydrogen Fuel Station Network Development," 2017, https://www.arb.ca.gov/msprog/zevprog/ab8/ab8_report_2017.pdf.
11. Neesc.org, "Northeast Electrochemical Energy Storage Cluster, 2017, [online] Available at: http://neesc.org/news/.
12. IHS Automotive, "Global Hydrogen Fuel Cell Electric Vehicle Market," [online] Available at: http://on.ihs.com/1SRkYJw.
13. E4Tech, "The Fuel Cell Industry Review," 2018, http://www.fuelcellindustryreview.com/.
14. "Riversimple Launches First UK Trial of Rasa Hydrogen Fuel Cell Cars," *Fuel Cells Bulletin* 2016(7):2, 2016, doi:10.1016/S1464-2859(16)30164-X.
15. Microcab website, http://www.microcab.co.uk/.
16. "GreenGT H2 Fuel Cell Racing Car Continues Development in France," *Fuel Cells Bulletin* 2015(7):4, July 2015, doi:10.1016/S1464-2859(15)30178-4.
17. "H2Bus Consortium to deploy 1000 Fuel Cell Buses across Europe," *Fuel Cells Bulletin* 2019(7):2-3, July 2019, doi:10.1016/S1464-2859(19)30267-6.
18. "California Approves New Flyer Fuel Cell Bus with Hydrogenics Unit," *Fuel Cells Bulletin* 2015(9):2-3, September 2015, doi:10.1016/S1464-2859(15)30240-6.
19. Liu, Z. et al., "China Progress on Renewable Energy Vehicles: Fuel Cells, Hydrogen and Battery Hybrid Vehicles," 2018, https://www.mdpi.com/1996-1073/12/1/54.
20. "Korean Joint Venture HyNet Aims to Set Up 100 Stations by 2022," *Fuel Cells Bulletin* 2019(4):8-9, April 2019, doi:10.1016/S1464-2859(19)30151-8.
21. Staffell, I., Scamman, D., Abad, A.V., Balcombe, P. et al., "The Role of Hydrogen and Fuel Cells in the Global Energy System," *Energy Environ. Sci.* 12:463-491, 2019, doi:10.1039/C8EE01157E.
22. "Nikola Launches Tre hydrogen-Electric Truck for European Market," *Fuel Cells Bulletin* 2018(12):4, December 2018, doi:10.1016/S1464-2859(18)30444-9.
23. "PowerCell Deal with Bosch for Automotive S3 Stack, Supplier 3M," *Fuel Cells Bulletin* 2019(1):12, January 2019, doi:10.1016/S1464-2859(19)30028-8.
24. "Toyota and Kenworth Collaborate on Zero-Emission Heavy Trucks," *Fuel Cells Bulletin* 2019(2):2, February 2019, doi:10.1016/S1464-2859(19)30041-0.
25. "Toyota and Kenworth Unveil Fuel Cell Heavy Truck at Port of LA," *Fuel Cells Bulletin* 2019(5):3-4, May 2019, doi:10.1016/S1464-2859(19)30180-4.
26. "Hyundai Motor, H2 Energy JV to Deliver 1600 Trucks in Europe," *Fuel Cells Bulletin* 2019(5):4, May 2019, doi:10.1016/S1464-2859(19)30182-8.
27. "Ballard Fuel Cells for California UPS Trucks in CARB-Funded Trial," *Fuel Cells Bulletin* 2018(11):2, November 2018, doi:10.1016/S1464-2859(18)30395-X.
28. "DHL, StreetScooter Develop Delivery Van with Plug Power Tech," *Fuel Cells Bulletin* 2019(6):2, June 2019, doi:10.1016/S1464-2859(19)30222-6.
29. "7-Eleven Japan and Toyota will Launch Joint Convenience Store Project in Autumn 2019," *Fuel Cells Bulletin* 2018(7):12-13, July 2018, doi:10.1016/S1464-2859(18)30260-8.
30. U.S. Department of Energy, "Fact of the Month November 2018: There Are Now More Than 20,000 Hydrogen Fuel Cell Forklifts in Use Across the United States," November 2018, https://www.energy.gov/eere/fuelcells/fact-month-november-2018-there-are-now-more-20000-hydrogen-fuel-cell-forklifts-use.
31. "Carrefour Showcases Hydrogen Station, Fuel Cell Forklift Fleet at New French Logistics Centre," *Fuel Cells Bulletin* 2018(12):4, December 2018, doi:10.1016/S1464-2859(18)30445-0.
32. "Plug Power, Linde MH Accelerate Hydrogen in Euro Manufacturing," *Fuel Cells Bulletin* 2018(7):12, July 2018, doi:10.1016/S1464-2859(18)30258-X.
33. Ministerial Council on Renewable Energy, Hydrogen and Related Issues, "Basic Hydrogen Strategy," Provisional Translation, 2017, https://www.meti.go.jp/english/press/2017/pdf/1226_003b.pdf, viewed August 21, 2019.
34. Chaben, J., "Japan Fuel Cell Developments," Fuel Cell & Hydrogen Energy Association, http://www.fchea.org/in-transition/2019/3/11/japanfuel-cell-developments, viewed October 12, 2019.
35. IEA, "The Future of Hydrogen," 2019, https://www.iea.org/hydrogen2019/.
36. Yan Xing, L., Hino Ryutaro, *Nuclear Hydrogen Production Handbook* (2011), https://www.crcpress.com/Nuclear-Hydrogen-Production-Handbook/Yan-Hino/p/book/9781138074682.
37. Simpson, A.P. and Lutz, A.E., "Exergy Analysis of Hydrogen Production via Steam Methane Reforming," *International Journal of Hydrogen Energy* 32(18):4811-4820, 2007.

38. Marculescu, C., Badea, A., Dinca, C., and Gheorghe, C., "Biomass to Syngas Conversion by Pyro-Gasification Process," in *Atti della conferenza "3rd WSEAS Int. Conf. on Waste Management, Water Pollution, Air Pollution, Indoor Climate,"* 978-960, https://www.semanticscholar.org/paper/Biomass-to-syngas-conversion-by-pyro-gasification-M%C4%83rculescu-Badea/5323f43816640d208ba0166c826c95b5f0e7ccae.
39. Concordblueenergy.com, "Concord Blue," 2019, [online] Available at: http://www.concordblueenergy.com.
40. Fargere, A., Kolodziejczyk, B., Carton, J.G., Lapeña, M.L., Pica, T.A., Karaca, C., Chae, Y., and Fuselli, L., "Hydrogen an Enabler of the Grand Transition," FEL-WEC, 2018.
41. Moro, A. and Lonza, L., "Electricity Carbon Intensity in European Member States: Impacts on GHG Emissions of Electric Vehicles," *Transportation Research Part D: Transport and Environment* 64:5-14, 2018. https://www.sciencedirect.com/science/article/pii/S1361920916307933.
42. Europa.eu, 2019, Europa [online] Available at: https://ec.europa.eu/clima/policies/transport/vehicles/cars_en.
43. IEA.org, 2019, IEA [online] Available at: https://www.iea.org/newsroom/energysnapshots/co2-intensity-of-new-power-gen-capacity.html.
44. Certifhy.org, 2016, EU Commission [online] Available at: https://www.certifhy.eu/.
45. Blanc, P., Zafar, M., Martino, F.D., Fargère, A., Carton, J., Kolodziejczyk, B., "Innovation Insights Brief New Hydrogen Economy - Hope or Hype?," World Energy Council, 2019.
46. Creti, A., Kotelnikova, A. (Fargère), Meunier, G., Ponssard, J.-P., "A Cost Benefit Analysis of Fuel Cell Electric Vehicles," 2015, https://hal.archives-ouvertes.fr/hal-01116997/document.
47. McKinsey, "A Portfolio of Power-Trains for Europe: A Fact-Based Analysis - The Role of Battery Electric Vehicles, Plug-In Hybrids and Fuel Cell Electric Vehicles," 2010, http://www.iphe.net/docs/Resources/Power_trains_for_Europe.pdf.
48. Kotelnikova, A. (Fargère), "Analysis of a Hydrogen-Based Transport System and the Role of Public Policy in the Transition to a Decarbonized Economy," 2016, Ph.D. thesis.
49. Stromberg, J., "What Really Sparked the Hindenburg Disaster?" *Smithsonian Magazine*, 2012. https://www.smithsonianmag.com/science-nature/what-really-sparked-the-hindenburg-disaster-85867521/.
50. Caton, P., "Water the Solution for Fuel Woes," *Tweed Daily News*, 2009, https://www.tweeddailynews.com.au/news/water-the-solution-floated-for-fuel-woes/285302/.
51. Sang-Hun, C., "North Korea Claims to Have Developed a Missile-Ready Hydrogen Bomb," *The New York Times*, 2017. https://www.nytimes.com/2017/09/02/world/asia/north-korea-hydrogen-bomb-missile.html.
52. Kolodziejczyk, B. and Ong, W.-L., "Hydrogen Power Is Here to Stay. How Do We Convince the Public That It's Safe?" *World Economic Forum.*, 2019. https://www.weforum.org/agenda/2019/04/why-don-t-the-public-see-hydrogen-as-a-safe-energy-source/.
53. Cardella, U. et al., "Economically Viable Large-Scale Hydrogen Liquefaction," *IOP Conf. Ser. Mater. Sci. Eng.* 171:012013, 2017. https://iopscience.iop.org/article/10.1088/1757-899X/171/1/012013/pdf.
54. SAE International Surface Vehicle Recommended Practice, "Recommended Practice for General Fuel Cell Vehicle Safety," SAE Standard J2578, Rev. Jan. 2009.
55. BloombergNEF, "Hydrogen: The Economics of Storage. Storing Clean Molecules at Scale, 2019.

Contact Information

EDGEresearch@sae.org.

The authors of this document together with the SAE Team responsible for its creation join in expressing our deepest appreciation to all of the individuals who contributed.

www.ingramcontent.com/pod-product-compliance
Lightning Source LLC
LaVergne TN
LVHW070941210926
848482LV00003B/271
* 9 7 8 1 4 6 8 6 0 1 0 1 5 *